Dr. Stepher
Morgan Co
1616 Wildle
Columbia, 9
phone: (803
Email: SLM

Construction and Assessment of Classification Rules

Construction and Assessment of Classification Rules

D. J. Hand
Open University, UK

JOHN WILEY & SONS

Chichester . New York . Weinheim . Brisbane . Singapore . Toronto

Other Wiley Editorial Offices

John Wiley & Sons, Inc., 605 Third Avenue,
New York, NY 10158-0012, USA

VCH Verlagsgesellschaft mbH, Pappelallee 3,
D-69469 Weinheim, Germany

Jacaranda Wiley Ltd, 33 Park Road, Milton,
Queensland 4064, Australia

John Wiley & Sons (Asia) Pte Ltd, 2 Clementi Loop #02-01,
Jin Xing Distripark, Singapore 129809

John Wiley & Sons (Canada) Ltd, 22 Worcester Road,
Rexdale, Ontario M9W 1L1, Canada

British Library Cataloguing in Publication Data

A catalogue record for this book is available from the British Library

ISBN 0 471 96583 9

Typeset in 10/12pt Times from the author's disks by Pure Tech India Ltd, Pondicherry
Printed and bound by Antony Rowe Ltd, Eastbourne

To Shelley

Contents

Preface

I regard building allocation rules, supervised pattern recognition, discriminant analysis, whatever name you like to use for the activity, as a *paradigmatic statistical problem*. By this I mean that it includes in its own microcosm (though nowadays a rather large microcosm!) all of the major issues of statistics: problem formulation, model building, estimation, uncertainty, prediction, interpretation, etc. But even that is too restrictive a description of the area. Major research efforts in recent decades mean that one might equally regard the subject as an area of computer science; certainly its links with that discipline are just as strong as they are with statistics. It is thus very much an interdisciplinary area. As such it benefits from the different philosophical and technological traditions in the two areas. They have their slightly different emphases and objectives, and the parallel research strands in the two sciences reflect this. When they come together, in supervised classification problems, some very exciting synergy results. I hope that this book conveys some of the richness and excitement of the field.

This book was originally intended to be a second edition of *Discrimination and Classification* (D. J. Hand, Wiley, 1981). However, in recognition of the fact that there has been so much progress since that book was published, I decided that a completely new book was more appropriate.

The field of classification and allocation is now huge, far too big to write a comprehensive book covering all publications; even McLachlan (1992) with his 60 pages of over 1200 references, acknowledges that 'by now there is an enormous literature on discriminant analysis, and so it is not possible to provide an exhaustive bibliography.' And he does not even attempt to cover neural networks. Moreover, there are also several recent books of substantial length which present in-depth examinations of subsets of the areas to be addressed; some of these are described at the end of Chapter 1. My aim was not to write a direct competitor; instead, I aimed to write a book which presented the central issues briefly and accessibly, was short enough to introduce the ideas without requiring a huge investment of time and which placed particular emphasis on comparison, assessment and how to match method to application.

Some readers may find it odd that there are so few numerical examples. It seems to me that numerical examples serve two purposes: they illustrate the steps in algorithms and they show examples of the results of an analysis (perhaps comparative results). Since this book does not discuss algorithms in depth, there seemed little point in showing steps in their arithmetic; essentially I have assumed that the reader who applies the methods described here will use computer programs which already implement suitable algorithms. For those who need to know about such things, guidance is given in the further reading. And as for showing the results of analyses, there are issues concerning what results it is useful to show for classification methods. In two dimensions one can show the decision surfaces resulting from different methods, but there are few general conclusions one can draw from such displays. In any case, two-dimensional problems are rare and such displays are impossible in more than two dimensions. One can always produce confusion matrices (or similar matrices for other measures of performance), but by concentrating on particular problems and data sets, their value is limited when it comes to drawing general conclusions.

Instead of numerical examples, I present some detailed discussions of real application areas, discussion of unusual classification problems and a detailed discussion of the relative merits of the different types of methods. I hope that this information will be more useful to a potential user of the methods than numerical examples applied to a handful of real data sets.

In general, the philosophy of the book is that different classifiers suit different problems. It is not an accident, or a ruse by researchers intent on attracting more funds and kudos, that so many qualitatively distinct kinds of classification rule have been invented. They have been invented because different problems demand different solutions.

I am indebted to several friends and colleagues for their helpful comments on various drafts of portions of this book. In particular, I would like to thank Andrew Webb, Niall Adams and Keming Yu. Not only did their comments improve the flow of the text, but they also prevented me from making too many mistakes. Of course, those mistakes which remain are entirely my own responsibility.

David J. Hand

The Open University
1996

Part I

Basic Ideas

CHAPTER 1

Introduction

1.1 THE PROBLEM

Researchers confronted with a classification problem for the first time sometimes ask, Which is the best type of classification rule? Presumably they are seeking an answer such as nearest neighbour methods, or neural networks, or logistic regression. However, the question is misleading: different problems have different features, and what is best for one problem is not best for another. What the researchers should really be asking is, Which is the best type of classification rule for my problem? Moreover, it is also necessary to say what is meant by *best*; what it is that the researchers are really interested in, and what it is that they want to optimise. These two issues—that different problems have different features and that what is regarded as good varies from problem to problem—are central to the theme of this book. This book thus emphasises properties of methods rather than algorithms for methods. Users, as opposed to developers, of classification systems do not need to concern themselves with the details of algorithmic manipulations. That is what computers are for.

The book is divided into four parts. The first, consisting solely of this chapter, introduces the basic concepts and ideas of supervised classification. Part II summarises methods for classification rule construction, covering classical statistical methods, more flexible methods such as neural networks, tree methods and nonparametric smoothing methods. Part III explores evaluation and assessment of classification rules. This is an important and underrated topic. All too often, rule performance is measured on the basis of the proportion of objects the rule misclassifies but, we argue, this is seldom actually appropriate. Different measures suit different problems. Finally, Part IV examines classification rules in practice, looking at some unusual problems and some important areas of application, and discussing the relative merits and demerits of the different types of rules under different circumstances.

Section 1.2 introduces the basic ideas and terminology and previews the content of the rest of the book. Section 1.3 provides an abstract overview of the problem. Section 1.4 gives some guidelines to further general reading.

1.2 BASIC IDEAS AND TERMINOLOGY

The word *classification* is used in various ways, so it is as well to be clear about the usage in this book right from the start. We can distinguish between *unsupervised classification* (or cluster analysis or unsupervised pattern recognition) and *supervised classification* (or supervised pattern recognition). Unsupervised classification refers to the process of defining classes of objects. That is, we are presented with a collection of objects and the aim is to formulate a class structure: to decide how many classes there are and which of the objects in the collection belong to which classes. In supervised classification, on the other hand, the class structure is known a priori and the aim is to formulate rules which allow one to allocate new objects to their appropriate classes. To illustrate, unsupervised classification will be used to discover if there are different types of depression, whereas supervised classification will be used to allocate a new person to one of the prespecified depression classes (i.e. diagnose them). Sometimes the single word *classification* is used for both supervised and unsupervised problems, but the context usually makes clear which usage is intended.

Unsupervised classification can, in fact, be divided into two types, according to the objective. One type seeks to identify naturally occurring structures in nature — to 'carve nature at the joints'. The search for different kinds of depression mentioned above would fit naturally into this mould. In contrast, the other type (sometimes called partitioning or dissection) simply seeks convenient divisions of the collection of objects. An illustration of this comes from marketing applications, where the population of potential purchasers is divided up into different types, not because of any belief that they are qualitatively different, but simply because it leads to a more efficient marketing strategy. The particular method of unsupervised classification one adopts will depend on which of these two aims one has.

This book is concerned with supervised classification. Material on unsupervised classification can be found in Section 1.4.

Problems of supervised classification occur in many different situations and many different guises. In view of this, it is hardly surprising that they also occur under different names. Problems of diagnosis, (pattern) recognition, identification, assignment and allocation are essentially supervised classification problems. In each case the aim is to classify an object into one of a prespecified set of classes. It also follows that the names will not always be used consistently by different authors. We have already commented above about the use of *classification* to refer to both supervised and unsupervised classification under different circumstances. Pankhurst (1991, p.1), for example, distinguishes between *identification* and *classification*, 'For a biologist, identification usually means finding the name for a specimen of animal or plant, and the specimen to be identified is usually assigned to a species.' So, to him, supervised classification is identification. He then goes on to say, 'Whatever sort of object is in question, it cannot be identified unless there is already a classification of like objects with which the new object can be compared.' So here the word *classification* is being used for the result of an unsupervised classification exercise: the set of classes which have resulted.

The archetypal supervised classification problem may be described as follows. Each object is described in terms of a vector of features (often numerical such as length, weight, IQ or temperature; but also possibly nominal such as colour, presence or absence of a characteristic, etc.). These features together span a multivariate space termed the *measurement space* or *feature space*. All objects have a true class, but this is typically unknown — the aim of the exercise is to formulate a rule for assigning objects to their true class. However, the true classes are known for a sample of objects, called the *design set, training set* or *learning set*. Thus, for the design set, we know both the feature vectors and the true classes. Our aim, then, is to use this sample to construct a general rule which can be used to predict the classes of new objects, based solely on the measurement vectors of these new objects. And this, at last, is how the word *supervised* gets into the process. The design set objects have known classifications: their true classes have been determined by a *teacher* or *supervisor*. In unsupervised classification, in contrast, there is no supervisor telling us the true classes: it is up to us to define the true class structure.

It is legitimate to ask what is the point of this exercise? If one can determine the true class for the design set, cannot one determine it for all objects? In fact there are many situations where this is impracticable. Prognostic situations are one type of situation; here we want to identify the class of an object before conventional measures of true class become possible. For example, if a definitive medical diagnosis depends on a post-mortem examination, we may want a classification rule which can be applied earlier, on the basis merely of some symptom patterns. The same applies in banking applications, where we want to identify good risk applicants before making the loan. Other situations are related to the objective. In machine recognition of speech, although it is possible for a human listener to allocate a spoken word to its correct class, the whole point of the exercise is to cut the human out of the loop: to build a machine which can do it unaided. Yet other situations relate to improved accuracy of machine recognition, reduced cost, greater speed, hostile environments, and so on. In each case we cannot use the true class, as determined by the 'teacher'.

The vector of measurements for a particular object corresponds to a point in the measurement space. Our aim is to decide to which class an object with that measurement vector should be assigned. This is done by using the design set to partition the measurement space into regions such that points falling in one region correspond to one class and points in another region correspond to another class. Clearly the crucial aspects of this partition are the (hyper)surfaces separating the regions. These are called *decision surfaces* since moving across such a surface changes the classification decision. Some methods for constructing classification rules focus solely on these decision surfaces.

If the design set is a random sample from the overall population, then regions of the measurement space densely populated by points corresponding to objects from class k (say), and only sparsely populated by points corresponding to objects from the other classes, should lead to new objects in that region being classified into class k. Thus in some sense our aim is to estimate $f(j|\mathbf{x})$, by estimates $\hat{f}(j|\mathbf{x})$, and assign a new object to the class corresponding to the largest estimate. Here j is the index of class identifiers and $\mathbf{x}$ is the vector of measurements, so that $f(j|\mathbf{x})$ is the probability that an object with measurement vector $\mathbf{x}$ will belong to class j. Whole classes of methods for supervised classification are based on direct estimation of the $f(j|\mathbf{x})$. Often particular distributional forms are assumed for these functions (which is fine if the assumptions are correct but not so good if they are not). In such cases, analytical manipulation of the algebra can often lead to substantial simplification of the resulting rules. Alternatively, *nonparametric methods*, such as those outlined in Chapter 5, make no distributional assumptions, instead concentrating on the local vicinity of the point at which a classification is required.

More generally, if we are only interested in finding the largest of the $f(j|\mathbf{x})$ at a particular $\mathbf{x}$, any monotonic increasing transformation of the estimates $\hat{f}(j|\mathbf{x})$ will give the same results. This relaxation can lead to simplification, as we shall see. Also, since we want to compare (estimates of) the $f(j|\mathbf{x})$ to give a classification, we could shift the focus of our efforts away from the individual $f(j|\mathbf{x})$ to estimate comparisons of them directly. In the two-class case, for example, we could concentrate on the ratio $f(1|\mathbf{x})/f(2|\mathbf{x})$, distinguishing regions where this ratio is greater than 1 from regions where it is less than 1. The term *discriminant function* is used to describe such comparisons of the $f(j|\mathbf{x})$, since they allow one to discriminate between the classes. One important special case of this is classical *linear discriminant analysis*, outlined in Chapter 2. Here, for certain important classes of $f(j|\mathbf{x})$, a logarithm of the ratio leads to a simple linear decision surface.

As indicated above, many methods are based on direct estimation and comparison of the $f(j|\mathbf{x})$. Others, however, use Bayes' theorem to invert the problem, focusing attention on the $f(\mathbf{x}|j)$. Thus $f(j|\mathbf{x}) = f(\mathbf{x}|j)\pi_j / \sum_k f(\mathbf{x}|k)\pi_k$, where π_j is the prior probability of class j. This separates the estimation of the class conditional distributions from the priors, with the advantage that one can then take separate samples from each class.

We have already contrasted methods which make distributional assumptions about the underlying functions with those which make no such assumptions. This distinction remains valid whatever functions—decision surfaces, probability distributions, discriminant functions, etc.—we are talking about. If we regard the two classes of methods as the two extremes of a continuum, we must also consider an intermediary class: a class of methods which have a very large number of parameters (though they still have parameters which need to be estimated from the design set). The number of parameters in such models may increase with increasing size of the design set. Such models include neural networks (though there are other such models, as described in Chapter 3), which have recently received so much publicity.

Errors in predicted classifications are most likely to occur in parts of the measurement space where $\max_j f(j|\mathbf{x})$ is only a little greater than $\max_{k \neq j} f(k|\mathbf{x})$. That is, where the second most likely class is nearly as likely as the most likely class. It is in these regions where the fact that the $f(j|\mathbf{x})$ are estimated is most likely to lead to errors in the order of the estimates $\hat{f}(j|\mathbf{x})$. In regions where the $f(j|\mathbf{x})$ are very different, the order is likely to be correctly estimated. Taking this further, we might expect to find most errors occurring in the vicinity of the decision surfaces; the decision surface is the estimate of the surface where $f(j|\mathbf{x}) = f(k|\mathbf{x})$ for the two most likely classes, so the vicinity of the decision surface is the place where the essential variation in estimation is most likely to lead to inversion of the order. All of this means that one might justifiably place priority on accuracy of estimation in those regions where $f(j|\mathbf{x}) \approx f(k|\mathbf{x})$ for the two most likely classes, or in the vicinity of the decision surfaces. On the other hand, it is in precisely these regions that the misclassification rate cannot be improved. For simplicity, consider the case of just two classes. Then in a region where $f(1|\mathbf{x}) \approx f(2|\mathbf{x})$ about 50% of the points will be misclassified, no matter which classification is chosen. Correct classification may mean that 51% are correctly classified and 49% incorrectly classified, but the converse is not a great deal worse.

Implicit in all of the above is an assumption that we are merely interested in minimising the proportion of objects misclassified — the *error rate* or *misclassification rate*. Although this is by far the most popular measure, and certainly the one most commonly adopted for methodological comparative studies of classification methods, the fact is that it is rarely what users of classification methods really want. Some of the weaknesses of simple error rate are as follows.

Error rate treats all the different types of misclassification as equally severe: a misclassification of a class A object into class B is regarded as equally serious as the converse. This assumes a basic symmetry of the class structure which rarely exists; certain types of misclassification are often more serious than other types (e.g. the relative severity of misclassifications when screening for a disease which can be easily treated if detected early enough, but which is otherwise fatal). If the relative severity of misclassifications can be quantified in terms of *costs* or *utilities*, it can be taken into account when constructing the classification rules. One can choose the decision surface to minimise the expected cost (or risk) of future misclassifications, rather than the expected number. (The decision surfaces corresponding to minimising the misclassification rate will arise when the costs of the different types of misclassification are assumed to be equal.)

The *confusion matrix* (Section 6.9) of a classification rule is the matrix obtained by cross-classifying the predicted class of objects from the population by their true class. This is often expressed in terms of proportions, so that the sum of the elements is unity. Various aspects of the performance of a classification rule can be obtained from this matrix. In particular, the sum of the proportions falling off the leading diagonal gives the error rate. With the elements of the matrix weighted by a corresponding cost matrix giving the costs of the different kinds of misclassification (so possibly an asymmetric matrix), the sum of the off-diagonal proportions gives the expected cost of the rule. It is common to assume that correct classifications incur no cost (so that, in particular, the diagonals of the cost matrix are zero) but this is seldom true in reality: making a classification involves expenditure, whether it turns out to be wrong or right. Of course, this might mean adding a constant positive term to all cells of the cost matrix; since this would make no difference, we could ignore it. However, one can also imagine a situation in which distinguishing between classes within different subgroups incurs costs which depend on the subgroup.

All this is all very well if one knows the costs associated with the different types of misclassifications, but all too often appropriate costs are difficult to determine. Various strategies have been developed to cope with this problem, ranging from ways of estimating costs based on the bets people are prepared to make through to the development of rules which yield the same result over ranges of cost values.

Taking account of costs is one important way of generalising simple error rate to make it a more realistic and useful measure of classification rule performance. But even this generalisation does not tap all aspects of performance. It does not, for example, tell us how much confidence we can place in the classifications. If, for example, a rule asserts that 70% of objects with a given measurement vector belong to class A, can we be confident that this is the case? With equal costs, we would then assign all objects with such a measurement vector to class A and expect to get 30% of them wrong. But what if the truth was that 51% of such objects belonged to class A? We would still be assigning the majority of the objects to the correct class, but (a) we should have nowhere near the confidence we seem to have for each classification and (b) we would be misclassifying far more than we expect, even though it is not possible to do better. On the other hand, what if 99% of such objects really belonged to class A? Again we are still classifying the majority to the correct class, but now 99% rather than the 70% we expect. And what if the costs are not determined with any accuracy, so that using a sharp threshold with the probability estimates is dubious? Then, ideally, we would like accurate probability estimates over a range of values, not merely near one particular threshold. These sorts of issues, and how to measure the different aspects of classification rule performance that they describe, are outlined in Chapter 6.

For a given classification problem and a given measurement space, the error rate is bounded below by what is called the *Bayes error rate*, discussed in Chapter 7. Typically this lower bound will be greater than zero. This is a natural consequence of the 'overlap' between the probability distributions of the different classes—a consequence of the fact that there exist measurement vectors $\mathbf{x}$ corresponding to members of more than one class. Only very rarely do situations arise when the classes do not overlap. This overlap has always been regarded as a fundamental aspect of statistical approaches to classification problems but the same has not been true for computer science approaches where, right from the early days (e.g. the perceptron, see Section 3.3), emphasis was placed on 'separability'.

If there exist measurement vectors corresponding to more than one class of objects then, no matter how cleverly one develops a classification rule, one cannot reduce the error rate to zero. Or, at least, one cannot do so without using extra information. And this leads us to the *reject option*—a way of sidestepping, if not solving, the problem. A threshold is chosen so that only those classifications in which one is confident (as determined by the threshold) are accepted. Decisions on others—perhaps where the estimated largest probability of class membership is only a little larger than the next largest estimated probability— are deferred while more information is collected. By this means, in principle at least, one could make the final overall error rate of the rule as small as one liked. However in practice there are limits to how long one will want to defer a decision. Note that, to adopt this approach, one needs not merely simple classifications from the rules but also actual estimates of the proportions estimated to belong to each class at each $\mathbf{x}$. Again, for confidence in the results, we need to be confident in the estimated probabilities $\hat{f}(j|\mathbf{x})$, not merely the fact that we are assigning the majority of objects with measurement vector $\mathbf{x}$ to the correct class.

So far we have said nothing about the actual estimation of error rates or other measures of performance, but this is a crucial aspect and needs to be treated carefully. There are, in fact, different kinds of error rate and different ways of estimating them. It is important to be sure that one is discussing the right thing. An important distinction is between the performance of a particular classification rule, developed on a particular design set, and the performance of a *method* of constructing classification rules. The distinction is the same as that between the accuracy of an estimate and the accuracy of an estimator. For a particular situation, in which one is presented with a particular design set from which one constructs a particular classification rule, one will want to know how well *this particular* rule performs. In contrast, before the rule has been constructed, and when one is considering what type of rule to construct, one will want to know how well rules of each type are expected to perform (and how far from the expectation they are likely to deviate). The first problem is conditional, conditional on the particular design set; the second is unconditional. At different times we may want estimates of each type of performance measure.

The aim of performance evaluation is to estimate how well the classifier is likely to perform in classifying new objects presented to it. That is, we want to know how well the rule will *generalise* to future objects. Implicit in the greater part of the work on supervised classification is the assumption that the distribution from which future objects will arise is the same as the distribution from which the design sample was (randomly) selected. However, this is not always a reasonable assumption: in many applications things evolve over time. We term such evolution *population drift* and present several examples of the phenomenon.

Population drift poses a difficulty for performance evaluation, a difficulty which does not permit a ready and general solution, since the circumstances of the drift, and the extent to which it permits modelling, vary from problem to problem.

A second compounding difficulty with performance estimation is that we are typically in the situation of wanting both to design and evaluate a classification rule from a given finite design set (only occasionally, most often in comparative simulation studies based on artificially generated data or where large quantities of data are available, do we have independent test sets). Now the rule will, in some sense, be optimised on the design set. That is, the form of the rule, the parameters describing it, will have been chosen to optimise some measure of performance relative to the design set. (It would be perverse, to say the least, deliberately to choose a classification rule which did poorly on the design set!) It follows that using this measure as evaluated on the design set is likely to overestimate its likely future performance. A great deal of work, especially on error rate estimation, has been carried out to try to circumvent this problem. Techniques such as cross-validation, the jackknife and bootstrap methods, which have widespread statistical application, were developed on error rate estimation applications.

Implicit in the preceding paragraph is the use of performance measures to choose between rules; this is besides their use as predictors of likely future behaviour of the classification rule. Performance measures may also be used as criteria for parameter estimation, which is really a special case of rule choice. All such issues can be subsumed under the more general statistical notion of *model selection*. Ideally, and in principle, one should choose the criterion through which a rule will be chosen to match the problem. For example, if minimising risk with a specified cost matrix is the real objective of the exercise then that is the performance measure which should be used to choose the rule. However, there are other factors which must be taken into account. An important factor is the problem of *overfitting*, in which the rule performs very well on the design set but not so well on future observations (and it is future observations we are interested in—we already know the true classes of the design set elements). This problem is discussed in the next section. Also, when a performance measure is being used to choose between rules, a premium is placed on ease of computation. If millions of models need to be examined in a search for the 'best' then it is important that the measure of quality should be quick to calculate. In the pattern recognition community, especially, this has led to much exploration of *separability measures*, measuring the difference between the class conditional distributions. Such measures are outlined in Chapter 6.

Although most of this book applies to the multiclass situation, some aspects of the special case of two classes are occasionally of interest. This is the most important special case, partly because problems are often defined in terms of the class of objects possessing some property and the complementary class of objects not possessing the property (so that two class problems are very common), and partly because other problems can often be reduced to a combination of two-class problems. However, there is also another special case which deserves mention: the one-class case.

The one-class classification problem arises when one has a design sample from the class in question, but no sample from other classes. The aim is to formulate a rule which will permit one to decide if a new object should be regarded as a member of the class in question. Such rules can be based on outlier detection methods. For example, if one can make a reasonable assumption about the form of the distribution from which the data arose, then *atypicality indices* can be based on the probability contours of the distribution. For a given point, its index will be the probability content of the contour through that point. Such an index will take values between 0 and 1, 1 signifying a high degree of atypicality. The set of classes is clearly defined in many classification problems, but this may not be true in others. For example, in many medical situations one is forced to create a class 'other', corresponding to diseases not included in the design set. In such a situation, atypicality indices can be used to determine whether it is appropriate to classify an object into one of the classes represented in the design set or whether it should be classified as 'other'. One strategy for doing this is to measure the typicality of the new object to each class, and classify it as 'other' if it is atypical of them all.

1.3 GENERALISATION

Our aim is to predict the true class j of an object for which we have only the measurement vector $\mathbf{x}$. We noted above that we can distinguish two situations, though we will almost always be concerned with just one of them. (Since, in a sense, the one we will be less concerned with is merely a degenerate special case of the other, this causes no difficulties.) The first of these situations arises when the classes are separable; when, for all measurement vectors $\mathbf{x}$, all objects with vector $\mathbf{x}$ have the same class. The second situation arises when there exists $\mathbf{x}$ such that not all the objects with measurement vector $\mathbf{x}$ belong to the same class (and typically, in practice, this will be the case for all $\mathbf{x}$). The first situation is sometimes described as a *noiseless* or *degenerate* situation. The problem is then one of generalising the known class memberships at the design set $\mathbf{x}$s to other points. That is, the problem is essentially one of *interpolation* or *extrapolation*.

Even when interpolating we can, in general, find an infinite number of models which will do the job. For example, if we are given a particular model m which fits the design set perfectly, then any model M which includes m as a special case will also be able to fit the design set perfectly. (For example, if coefficients for a quadratic model can be found so as to fit the design data perfectly, then coefficients for a higher-degree model can always be found which will predict the design data perfectly.) Such generalisations of m will typically not agree with each other at points of the measurement space other than the points of the design set, so we have the problem of how to choose between such models.

As in Section 1.2, let $f(j|\mathbf{x})$ denote the probability that an object with measurement vector $\mathbf{x}$ belongs to class j. The first situation arises when $f(j|\mathbf{x})$ is a delta function, taking the value 0 for all classes except one, for which it takes the value 1, at each $\mathbf{x}$. One way of considering of the second situation is to regard it as arising when the first situation is contaminated by noise. Thus, instead of the $f(j|\mathbf{x})$ consistently taking the value 0 or 1 at each $\mathbf{x}$, the extra noise drags the function away from 0 and 1 so that, for a given $\mathbf{x}$, each of the js sometimes arises as the response. This may be a useful perspective under some circumstances, but a more useful perspective for the purposes of this book is to regard $f(j|\mathbf{x})$, right from the start, as being a probability which is to be modelled. Thus, all classes of objects can occur at a given $\mathbf{x}$, and our aim is to model the *probability* of their occurrence. Using this estimated probability we can produce a classification. (Although, for crude classification purposes, we are interested solely in the decision surface, which is derived from the $f(j|\mathbf{x})$, we have already noted above that it is convenient and more generally applicable to discuss things in terms of the $f(j|\mathbf{x})$.)

So, whereas in the first situation the classes are perfectly separable (if by a potentially very complex decision surface), in the second (for a given set of measurement variables) they are not. No matter how clever or convoluted the decision surface in the second case, since there exist $\mathbf{x}$s at which objects from both classes occur, the classes cannot be perfectly separated. *Most real problems are of this second kind.* The first situation allows us to obtain a zero future error rate, at least in principle. But we cannot do this in the second situation, no matter how hard we try. Of course, even in the second situation there is some error rate e which is minimal for a given set of measured predictor variables. This is the *Bayes error rate* mentioned above and discussed in Chapter 7. It is the error rate produced by a rule based on the true probabilities $f(j|\mathbf{x})$. And, again in principle, we can achieve this. Note also that, by changing the set of variables which is measured, we can change the Bayes error rate. By astute choice of variables and assuming that we have a large enough design set to derive accurate classification rules, we can in principle make the error rate as small as we like (subject to limitations imposed by irreducible quantum variation).

The phrase 'in principle', qualifying the lower bounds on achievable performance, is important. It will only be achieved by classification rules for which the estimates $\hat{f}(j|\mathbf{x})$ exactly coincide with the 'true' values. (In fact, as we noted, for classification purposes we should really talk in terms of decision surfaces.) This corresponds to infinite design sets and it also introduces deep issues of whether such true values really exist. In particular, recalling the notion of population drift, no problems are static. Although one might imagine true $f(j|\mathbf{x})$ as existing at the time the design set was collected, it is the essence of most classification problems that there is potential for the underlying distributions to vary somewhat (not greatly, one hopes, or the whole enterprise is doomed) with time, with space, with variations in the sampling process, and so on. This means that slightly different true $f(j|\mathbf{x})$ will apply to the distributions from which new objects are chosen.

Our aim, then, is to predict the true class of an object at measurement vector $\mathbf{x}$, but we cannot do this because, for a given set of variables, objects of each class occur at $\mathbf{x}$. We thus modify our objective to being one of finding good estimates of the $f(j|\mathbf{x})$.

We want to find estimators which lead to estimates $\hat{f}(j|\mathbf{x})$ which are close to the true values $f(j|\mathbf{x})$. We would like to do this by choosing the (parameters of the) estimated functions so as to minimise some measure of the difference between the estimated and the true values. But we cannot do this because we don't know the true functions; if we did, we would not need to go through this rigmarole. However, we do have *information* about the true functions, namely, the information in the design sample. So somehow we replace 'true' by 'design set' in our measure of the difference between the true and estimated functions, then choose the parameters of the estimates so as to minimise this difference.

Obviously we want to use a family of estimators which are flexible enough to model the twists and turns of the true $f(j|\mathbf{x})$. On the other hand, since we will be choosing our estimate so as to minimise some measure of discrepancy between the design set and the estimate, we do not want the estimator to be too flexible. Otherwise we could end up predicting the design set very well, to the extent that we would also be modelling the idiosyncrasies of that particular finite sample, idiosyncrasies which fail to reflect aspects of the underlying true probabilities. We have already explained how the term *overfitting* is used to describe the matching of peculiarities of the design set that are not representative of the underlying true population distributions. Clearly the problem is a difficult one: we want to model all the peculiarities of the true functions and none of the extra peculiarities which are due to the particular way the (finite) design set has fallen. And the problem is that we do not know which aspects of the design data are specific to the sample and which represent underlying characteristics of the true functions. Very powerful and flexible model forms for the decision surface are likely to be counterproductive in that they will lead to overfitted estimates. We want something flexible enough to model the twists and turns of the unknown true decision surface but which does not go further and model the chance fluctuations due to the fact that the design set is only a finite sample.

Let us make the above ideas a little more concrete. For simplicity, suppose that we have just two classes and for notational convenience temporarily represent $f(1|\mathbf{x})$ by f. Consider a particular design set, denoted by D, and a particular measurement vector $\mathbf{x}$. Then one way of measuring the accuracy of the probability estimate $\hat{f} = \hat{f}(1|\mathbf{x})$ at $\mathbf{x}$ is using $E_{c|\mathbf{x}}(\hat{f} - c)^2$, where the expectation is over the true class c ($= 0$ or 1) at $\mathbf{x}$. Now,

$$E_{c|\mathbf{x}}(\hat{f} - c)^2 = E_{c|\mathbf{x}}(\hat{f} - E_{c|\mathbf{x}}c)^2 + E_{c|\mathbf{x}}(E_{c|\mathbf{x}}c - c)^2$$

If we now summarise this decomposition over the measurement space **x**, we obtain an overall measure of the performance of the rule, partitioned into two components. Note that here the last term is independent of the classification rule, so that it provides a lower bound on the accuracy which an estimate can obtain. In Chapter 6, where we examine performance measures in detail, the three terms above, when summarised over **x**, are respectively called *inaccuracy*, *imprecision* and *inseparability*.

This brings us back to another point already introduced in the preceding section. The above expression, when integrated over **x**, measures accuracy of an *estimate*. For a given design set, this will tell us how well the chosen method performs and is certainly something we will be interested in so that, for example, we can compare competing classification rules. In general, however, we also want to know about the accuracy of *estimators*. That is, we want to know whether an estimation procedure can be relied on to give good estimates, regardless of whether a particular instance turns out to be good. To explore this, we need to evaluate the expectation of the above over different design sets. The second term on the right-hand side of the above expression is invariant to design set changes, so we will focus on the first term. Still fixing attention on a given **x**, we have

$$E_D E_{c|\mathbf{x}} (\hat{f} - E_{c|\mathbf{x}} c)^2 = E_D (\hat{f} - E_D \hat{f})^2 + (E_D \hat{f} - E_{c|\mathbf{x}} c)^2$$

that is, a decomposition into the variance and the (squared) bias of the estimator. An estimator which is very flexible, in the sense that it can model an irregular $f(1|\mathbf{x}) = E_{c|\mathbf{x}} c$, will tend to have a high variance: it will also tend to fit the design set well, so that it will vary dramatically from design set to design set. Conversely, an estimator which is inflexible, and which does not change much between design sets, so that it has a small variance, could well be substantially biased unless **x** happens to be in a region where $E_D \hat{f}$ is approximately the same as $E_{c|\mathbf{x}} c$.

This decomposition into bias and variance provides the solution to an issue which was a source of confusion in the early days of pattern recognition and statistical classification work. Naively one might assume that the more complex the model, the greater its flexibility, so that it would be likely to do better in future classification tasks. However, we have already seen that the more complex the model, the better it fits the design set, and eventually it goes beyond fitting the underlying $f(j|\mathbf{x})$ and starts to fit the additional random variation. One manifestation of this in statistics is the fact that a simple linear discriminant analysis is often more effective than a quadratic discriminant analysis. Quadratic discriminant analysis is more flexible, so it will probably be able to provide a more realistic model for the unknown decision surface (which is unlikely to be exactly linear), but it also has a great many more parameters to be estimated and therefore a substantially higher variance. Another example lies in the number of variables spanning the measurement space. Again naively, one might expect that since adding more predictor variables can only add information about the separability between classes such a process can only lead to improved performance. However, after a certain point, adding more variables can be counter-productive, as the model begins to fit the random component in the design set. Further discussion of this relationship between dimensionality and misclassification rate is given in Hand (1981a, Chapter 6). The moral is that a simpler, more restricted model may be more effective.

Overall, to minimise the mean squared error, we ideally want to minimise both bias and variance. One way to do this is to take a very flexible rule, which will have low bias, and reduce its variance by increasing the size of the design set. For sufficiently flexible rules, such as neural networks (Chapter 3) and nearest neighbour methods (Chapter 5), it can be shown that the mean square error $E_D E_{c|\mathbf{x}} (\hat{f} - E_{c|\mathbf{x}} c)^2$ can be made arbitrarily small by increasing the model complexity while increasing the size of the design set sufficiently rapidly. In statistical terms, all we are saying is that such models are consistent. (Of course, this does not mean that the resulting classification rule will have zero error rate. This is prevented by the overlap of the probability distributions for the classes. What it means is that, by increasing the design set size and the model complexity appropriately, we can get arbitrarily close to the best achievable performance with the given set of measured variables.)

Asymptotic results with arbitrarily complex models are all very well, but in practice we have finite design sets to contend with. To obtain a good estimator in such cases we must somehow achieve a compromise between bias and variance, trading off one against the other so that the combination above is minimised. In the terminology of generalised linear models, our aim is to model the systematic part of the model but not the random part, and somehow we must extract the systematic part from the design data.

Sometimes we have extra information which may guide us to a good family of forms for the classification rule that suits the problem in question, a family flexible enough to fit the $f(j|\mathbf{x})$ without being overflexible. This information might come from prior knowledge, perhaps from problems similar to the one being studied (no problem ever exists in isolation). For example, we might know that the family of linear or quadratic decision surfaces is effective for problems of the type in question. It is also worth remarking that simple classification performance is just one criterion which needs to be considered when choosing a model; others are speed of classification, whether the structure of the model must permit ready interpretation, and so on. These matters are discussed in Chapter 11.

Several strategies have been proposed for choosing a suitable compromise between flexibility of the estimator and the danger of overfitting. The most obvious approach is simply to try to choose the number of parameters so that a good but not overfitted model is obtained. Examples are the choice of the number of nodes and connections in a neural network and the number of nodes in a tree classifier. To choose a model of near optimal complexity one will need to evaluate the model on data sets other than the set used to construct it and choose a model which appears to do well (or else one will overfit). But the model's performance on this second data set cannot then be taken as a measure of its likely future performance. The fact that the second data set has been used to select the model makes it part of the design process. The model's performance on this set will be optimistically biased as an estimate of future performance—hardly surprisingly since the model was chosen because it gave the best performance on this particular second data set. A third independent data set is needed to evaluate likely future performance. In fact, subtle data reuse methods, mentioned in Section 1.2, such as cross-validation and bootstrap methods, allow the calculations to be performed on just a single set. Some of them are discussed in Chapter 7 for the special case in which the performance measure is error rate.

To choose an effective model, one has to perform some kind of search through the space of models. Typically this space is vast, so a restricted search is necessary. Common procedures include forward and backward stepwise methods (and combinations of the two). In a forward procedure one begins with a simple model and adds terms to it, gradually making it more complex. At each step the added term is that which produces the most effective new model. In a backwards procedure one begins with a complicated model and gradually deletes terms from it, deleting that term which leads to least degradation in the model fit. Since only single, or at least simple, terms are considered for inclusion/exclusion at each stage, the search is restricted. In the forward case, for example, the possibility of including the combination of a series of terms in one step, along with all their interactions, is not considered. Forward and backward procedures do not necessarily lead to the same final model; this is because they are restricting the search through the model space, and restricting it in different ways. Combination approaches adopt a process such as adding two terms to the model at each step, then deleting one, adding two, and so on. Other search procedures have also been explored. For example, if the goodness of fit measure is monotonic in the number of terms in the model, then a branch and bound method can be used. Some of these restricted model search procedures are described for the particular problem of choosing variables in Section 9.1.

A second approach to trying to find a good compromise between bias and variance is to *regularise* the model by adding a *penalisation* term to the criterion measuring goodness of fit to the design data. Suppose that the goodness of fit criterion takes small values for a good fit. Then the penalisation term is designed to increase as the number of parameters increases. The overall criterion (goodness of fit plus penalisation term) is then minimised at some compromise between a highly accurate fit with a complicated model and a poor fit with a simple model. Clearly the choice of penalisation function and the relative weights accorded to it and the goodness of fit measure are crucial in obtaining good results. Various penalisation functions have been proposed, including

- The Akaike information criterion (AIC). Likelihood is a suitable measure of goodness of fit (though increasing rather than decreasing). The likelihood of the design set will increase as more parameters are added to a model, so, to prevent overfitting, Akaike suggested reducing the likelihood by an amount

$2p$, where p is the number of parameters in the model. This value is based on consideration of the difference between the model's fit to the design data and its fit to an independent test set. Several modifications and similar criteria have also been proposed, including C_p and Schwarz's criterion.

- A function which penalises rapid variation in the estimate. For example, we could base one on the second derivative of $f(1|x)$, modifying the sum of squared errors to yield $\sum(\hat{f}_i - c_i)^2 + \lambda \int (\partial^2 \hat{f}/\partial \mathbf{x}^2)\mathrm{d}\mathbf{x}$.

- The sum of squared weights in the model, $\sum w_i^2$ is a common regularisation term in neural networks. This process, called *weight decay*, penalises large weights, so forcing the model into the region of the logistic transformation which is almost linear (see Chapter 3). That is, it tends to flatten out the component decision surfaces of the network.

- Minimum message length and minimum description length. These methods balance a measure of fit of the model to the data with a measure of model complexity; both measures are defined in terms of the common currency of the length of a binary string. Although closely related to Bayesian methods in statistics, they have their roots in computer science.

A third way to find a bias/variance compromise is to smooth an overfitted function. One example of this approach is pruning and averaging trees, as described in Chapter 4. That is, one builds a large tree—possibly even going so far as to produce leaf nodes containing only single classes—then either truncates leaf nodes or averages several or many such trees. A second example of this approach is combining the result of several classifiers, which can be done in various ways. One can, for example, simply average the predictions (either the predicted classes or the predicted probabilities), take a majority vote, or use the predictions as input to a higher level classifier. These ideas are discussed in Section 9.6.

A fourth strategy for finding a suitable compromise is to smooth the *data* rather than the model. The problem is that a flexible model overfits the design set. If the design set is smoothed in some way, prior to the model fitting, the problems caused by overfitting will be reduced. One way to smooth the design set is to replace it by multiple copies, but where each copy is slightly randomly perturbed, and then fit this enlarged data set. This is analogous to a stochastic empirical variant of kernel smoothing (Chapter 5). As with penalisation, one has to decide how much random variation to add to each of the original data points. The variance of this new variation becomes a parameter in the model.

1.4 FURTHER READING

A general discussion of matching problems to solutions can be found in Hand (1994a). Further information on unsupervised classification can be found in books on cluster analysis such as Anderberg (1973), Everitt (1974), Gordon (1981) and Späth (1985), as well as chapters in more general books on multivariate statistics, such as Krzanowski and Marriott (1995, Volume 2, Chapter 10), and as chapters in books covering both supervised and unsupervised classification (e.g. Hand 1981a; Ripley, 1996). A discussion of unsupervised classification in the context of depression research is given in Dunn *et al.* (1993).

Dawid (1976) describes the direct use of $f(j|\mathbf{x})$ as the *diagnostic paradigm* and the indirect use, via Bayes' theorem, of $f(\mathbf{x}|j)$ as the *sampling paradigm*. The $f(j|\mathbf{x})$ are what we are really interested in, so one might argue that we should focus on methods which place them centre stage. More generally, however, in some situations one might regard the j as preceding the $\mathbf{x}$, in some sense, so that the $f(\mathbf{x}|j)$ are more natural (for example, if the values in $\mathbf{x}$ are symptoms resulting from a disease which is to be diagnosed), although this is not always appropriate (if, for example, the values in $\mathbf{x}$ are environmental causes of a disease).

For a particular point, Aitchison *et al.* (1977) define an atypicality index in terms of a normal distribution fitted to the design sample.

Akaike's criterion is introduced in Akaike (1973), the C_p statistic in Mallows (1973), and Schwarz's criterion in Schwarz (1978). The ideas behind minimum message length are described in Wallace and Freeman (1987) and Oliver and Hand (1994), and the related ideas behind minimum description length in Rissanen (1987, 1989).

At various points in this book we have noted the emphasis placed by the pattern recognition community on perfect separability between classes, at least in the early days. This is in contrast to the perspective of the statistical community, which took the viewpoint of class conditional distributions and always expected overlap between the classes. The overlap viewpoint is the more practical in most real problems, but work derived from the separability viewpoint has led to some interesting developments, especially in recent years. Early work is illustrated by explorations of the proportions of all the partitions of a data set which can be effected by a linear functions of the variables defining the measurement space (e.g. Cover, 1965). This has been very substantially generalised under the guise of *computational learning theory*: given a sample from an unknown true classifier, we seek a function which, with high probability, will provide us with a good approximation to the unknown true classification rule. The approaches are thus termed *probably almost correct (PAC) learning* (Valiant, 1984; Anthony and Biggs, 1992). A particular emphasis of this work is on how large a design set is needed in order to attain a specified probability of obtaining a result within a given bound of the true classification rule's performance. PAC learning is a very general theory, independent of the learning algorithm and the design set distribution. This is both a strength and a weakness — the results are dominated by worst case behaviour.

The large size of the parameter spaces of some of the models which have been developed recently (such as those described in Chapter 3) is reminiscent of the large size of configuration spaces in statistical physics. Since statistical physics is an area which has been very extensively explored, this raises the possibility that some of the ideas and methods developed there might be usefully applied in the machine learning context. In contrast to the PAC learning approach, the statistical physics approach describes the typical behaviour of the systems. The parameter estimation process is assumed to be stochastic, yielding a Gibbs distribution for the parameters. Details of such approaches are given in Hertz *et al.* (1991), Seung *et al.* (1992) and Watkin *et al.* (1993).

We commented in the preface that several other general books on the topics dealt with here had recently appeared. Each book has its own emphasis and some of the more important ones are briefly described below.

Hand (1981a) was an early introduction to the field, attempting to pull together the work of the statistical and pattern recognition communities. As remarked in the preface, the field has advanced substantially since its publication. This is largely due to the impact of the computer, which permitted the development of methods hitherto impractical. This has, in turn, stimulated development of the theory underlying the problems, concepts and methods of supervised classification.

Keinsoke Fukunaga has been one of the key researchers in statistical pattern recognition, making important advances in a number of areas. Fukunaga (1990) is the second edition of his original 1972 edition (a substantial expansion, from 369 to 591 pages) — one of the bibles of early pattern recognition researchers. In a sense the book is rather abstract, perhaps towards the opposite end of the spectrum from the present one, which emphasises real problems. The reference list is rather limited (for a comprehensive list, see McLachlan, 1992) but the book does contain computer projects and exercises, making it suitable for use as a text.

Hertz *et al.* (1991) provides one of the best early (in neural networks terms) introductions to the ideas underlying neural networks. It still stands the test of time, and if one wanted to approach things from the neural network perspective, this book and that of Bishop (1995) would make an excellent pair.

McLachlan (1992) is a *tour de force*. It provides a detailed comparative critical assessment of the statistical and pattern recognition work. This book is essential reading for serious researchers in the area, providing references to most of the important work outside of neural networks. It is a pity there is no discussion of neural networks, but since the book is already over 500 pages long, perhaps this would have made it unwieldy. It should also be remarked that pattern recognition and discriminant analysis are experiencing such a resurgence of interest that the four years since the appearance of McLachlan have seen much important work published (such as on recursive partitioning methods, sample reuse methods and generalisation).

Bishop (1995) is an excellent introductory text from the perspective of neural networks. It includes exercises and covers most of the main areas and topics which are currently attracting research interest.

Ripley (1996) is a recent publication which does a good job of unifying the work undertaken by the various communities studying classification problems, communities which have not always recognised each other's contribution. This can be beneficial, because it gives different views of a problem, and detrimental, since it inevitably leads to duplication and wasted effort. Ripley's book is comprehensive, including discussion of topics such as neural networks, PAC learning and belief networks as well as the standard statistical pattern recognition staples. The book also shows a depth of theoretical understanding of underlying principles which is not present in all books on the topic.

Part II

Constructing Rules

CHAPTER 2

Fisher's LDA and Other Methods Based on Covariance Matrices

2.1 INTRODUCTION

Fisher's method of linear discriminant analysis (LDA) has played a central role in allocation rule methodology. It can be derived as an ordinary least squares (OLS) regression, was the one of the earliest formal statistical methods to be developed and is still widely used some 60 years later. However, especially in recent decades (in order to tackle new kinds of problems and due to the possibilities provided by the computer), it has been extended in several ways. In view of this it seems quite appropriate to devote an entire chapter to LDA and its variants.

Perhaps the single most important extension of the basic method is quadratic discriminant analysis (QDA), which can be viewed in various ways. It is a way of including squared as well as linear functions of the predictors. This means that the decision surface, though still linear in the space of features, is quadratic in the original measurement space. Or it can be regarded as a way of allowing different class covariance matrices. This method is described in Section 2.4.

QDA requires the estimation of many parameters unless very few predictor variables are involved. Unless the sample size is correspondingly large, this will mean that the estimates have large variance. One might say that, although LDA runs the risk of underfitting the data, QDA risks overfitting it. Overfitting can be a problem even if the sample size is large relative to the number of variables. A common such situation, termed *multicollinearity*, arises when the predictor variables are highly correlated. In such a case, even a large design set will lie predominantly in a low-dimensional subspace of the entire predictor space and will produce instability in the estimates of the regression coefficients (large variance, dramatic shifts from design set to design set). An extreme case arises when there is a perfect linear relationship between some of the predictor variables. This means the covariance matrix is singular and cannot be inverted to give the necessary estimators. Such singularities also arise if the sample size n is less than $d+C$ (d variables and C classes). Various ways to tackle such problems have been proposed.

Sometimes one has some (prior, theoretical) idea about the relationships between the predictors—about the covariances between them. When this is the case one might be able to postulate a particular structure for their covariance matrix. This will be equivalent to reducing the number of parameters required. Methods based on this idea are outlined in Section 2.5, where we examine covariance matrices arising from random effects models, from covariance selection models, and common principal components models.

A second approach is simply to reduce the number of variables used, so that the covariance matrix is of smaller order. Variable reduction may be by *selection* of a subset of variables from those measured or by *extraction* of a small set of functions of the measured variables. Methods of selection and extraction are briefly reviewed in Section 9.1, but in Section 2.6 we look at methods which have been developed in the present context, starting with *principal components regression* (PCR) and *partial least squares* (PLS). Both of these methods extract the variables on the basis of the overall distribution of the samples in the predictor space. In contrast, the *SIMCA* and *DASCO* methods extract variables for each class separately, computing separate similarity measures from a new object to each class.

A third approach is based on the notion of *regularisation* (Section 2.7). *Ridge regression* (RR) takes the overall covariance matrix of the predictors and shrinks it towards the identity matrix. The aim is to reduce the mean square error in the estimates of the regression parameters, even though bias may be increased. *Regularised discriminant analysis* (RDA) shrinks each class's covariance matrix towards the average covariance matrix (so that it lies between LDA and QDA) then further shrinks the result towards the identity matrix.

Finally, although regularisation methods can be thought of as minimising a weighted combination of a conventional criterion (e.g. least squares fit) and a roughness penalty, the *shortest least squares* (SLS) approach, described in Section 2.8, finds the minimum least squares solution which minimises the length of the parameter vector. This matters when the sample size is so small relative to the number of variables that the covariance matrix is singular since then the least squares solution (using a generalised inverse) is not unique.

2.2 THE BASIC FISHER METHOD

Consider the special case of two classes. Our aim is to compare an estimate of $f(1|\mathbf{x})$ with a threshold and classify an object with measurement vector $\mathbf{x}$ to class 1 if the estimate is greater than some threshold and to class 0 otherwise (or, equivalently, to assign the object to the class corresponding to the larger of

$f(1|\mathbf{x})$ and $f(0|\mathbf{x})$). However, since the estimate of $f(1|\mathbf{x})$ is simply to be compared with a threshold, any monotonically related transformation of the estimate will be equally valid (with the threshold transformed in a corresponding way). In general, we need not be overly concerned with estimating $f(1|\mathbf{x})$ precisely, so long as we maintain its order property. In particular, if we believe that $f(1|\mathbf{x})$ increases monotonically (or approximately monotonically) in some direction in the $\mathbf{x}$ space, then distance from some origin, *measured in that direction*, will do. That is, if we believe this, then we can compare a simple linear function $\mathbf{x}'\mathbf{w}$ with a threshold. The problem then is to find a good estimate for the direction $\mathbf{w}$.

One obvious approach to estimating $\mathbf{w}$ is to choose that direction in which the two design set samples are most widely separated. Of course, this does not completely define things—points, not samples, have distances between them. A natural resolution to this problem is to summarise the samples by their means and to adopt the distance between the sample means in the direction. So, for a given direction $\mathbf{w}$, the separability between the two samples is defined as $\mathbf{w}'(\bar{\mathbf{x}}_1 - \bar{\mathbf{x}}_0) = \mathbf{w}'\bar{\mathbf{x}}_1 - \mathbf{w}'\bar{\mathbf{x}}_0$.

This is all very well, but it takes no account of the fact that the x_j will generally have different variances. Indeed, they may be measured on arbitrary noncommensurate scales, so that the relative sizes of the variances may be meaningless. Moreover, the simple distance measure above makes no allowance for the covariances between the measurements. Allowance for both of these issues can be made by adopting, instead, a distance measure which standardises for the variances and covariances within the populations, in the direction $\mathbf{w}$. In fact, to simplify things, the assumption is made that the two populations have equal covariance matrices (we will drop this restriction later). Then the estimated common within-class variance in the direction $\mathbf{w}$ is $\mathbf{w}'\mathbf{S}\mathbf{w}$, where $\mathbf{S}$ is the estimated within class covariance matrix. This leads to the distance measure (squared, now, for mathematical convenience; since squaring is a monotonic function, this makes no difference to comparisons of distances) $D(\mathbf{w}) = (\mathbf{w}'\bar{\mathbf{x}}_1 - \mathbf{w}'\bar{\mathbf{x}}_0)^2/\mathbf{w}'\mathbf{S}\mathbf{w}$. Note that, for a given direction $\mathbf{w}$, this is simply the square of the t-statistic for comparing two independent groups.

$D(\mathbf{w})$ gives the distance or separability between the two classes in the direction $\mathbf{w}$. To find the best direction we need to find that $\mathbf{w}$ which maximises $D(\mathbf{w})$. This is easily shown to be given by the direction $\mathbf{w} \propto \mathbf{S}^{-1}(\bar{\mathbf{x}}_1 - \bar{\mathbf{x}}_0)$. This can be regarded as an estimate of $\Sigma^{-1}(\mu_1 - \mu_0)$, the corresponding population direction, with Σ the assumed common within class covariance matrix.

It is important to note that no distributional assumptions have been made in the above derivation. All that has been done is to identify a direction which best separates members of the two classes in some sense. A new object, with measurement vector $\mathbf{x}$, will be classified according to its position on this continuum. If it is nearer to the class 0 objects, it will be classified as class 0; if nearer to class 1 objects, as class 1. The choice of threshold for the decision point will depend on the class priors and the relative costs of the two types of misclassification. We will return to this issue below.

Although no distributional assumptions have been made, the distributions were summarised in terms of their first- and second-order moments. If the underlying distributions could in fact be completely described by those moments, we would expect the resulting classification rule to be optimal. That is, if the populations of the two classes had distributions which were completely described by their mean vectors and their covariance matrices (with the additional constraint that the two classes had identical covariance matrices), the direction $\Sigma^{-1}(\mu_1 - \mu_0)$ would yield the optimum classification rule. Such distributions are termed *elliptical distributions* and by far the most important special case is the multivariate normal distribution. Let us examine it in more detail.

Suppose that the classes have multivariate normal distributions $N(\mu_1, \Sigma)$ and $N(\mu_0, \Sigma)$ with respective priors π_1 and π_0. Then

$$\frac{f(1|\mathbf{x})}{f(0|\mathbf{x})} = \frac{\dfrac{\pi_1}{(2\pi)^{d/2}|\Sigma|^{1/2}} \exp\left[-\dfrac{1}{2}(\mathbf{x} - \mu_1)'\Sigma^{-1}(\mathbf{x} - \mu_1)\right]}{\dfrac{\pi_0}{(2\pi)^{d/2}|\Sigma|^{1/2}} \exp\left[-\dfrac{1}{2}(\mathbf{x} - \mu_0)'\Sigma^{-1}(\mathbf{x} - \mu_0)\right]} \tag{2.1}$$

For these distributions an optimum classification can be obtained by comparing this ratio with a threshold. In particular, this threshold will be unity if the costs of the two types of misclassification are equal. However, things can be simplified. First note that the denominators in the two p.d.f.s cancel. Secondly, if we take logs (a monotonic transformation, therefore the resulting expression still needs only to be compared with a threshold, now zero if we still assume equal costs) equation (2.1) simplifies to

$$\mathbf{x}'\Sigma^{-1}(\mu_1 - \mu_0) + \ln(\pi_1/\pi_0) - \frac{1}{2}\mu_1'\Sigma^{-1}\mu_1 + \frac{1}{2}\mu_0'\Sigma^{-1}\mu_0$$

This is precisely the result claimed above: the classification rule will consist of comparing $\mathbf{x}'\Sigma^{-1}(\mu_1 - \mu_0)$ with a threshold (which depends on the μ_k, Σ, the priors and the costs). Plugging in the maximum likelihood estimators of the means and covariance matrix leads to the classification rule developed above.

To summarise, Fisher's linear discriminant method is optimal for elliptical distributions, such as the multivariate normal, with equal covariance matrices (but it does not 'assume' multivariate normal distributions, as is incorrectly asserted by many texts). On the other hand, the method may perform well even if the distributions are not elliptical. For example, there is evidence to suggest that it does well even for multivariate binary data when the true optimum decision surface is linear.

Since Fisher's method is based on second-degree terms in the xs, one might expect a relationship of some kind to regression, which minimises a sum of squares criterion and so is also based on such terms. In fact, the relationship is very close.

Again we consider two classes, with the class membership of each point being described by a variable y coded as 0 or 1. We can now use ordinary least squares regression to formulate a class membership prediction rule, classifying a new case as class 0 if the prediction from the regression $\hat{y}$ is less than some threshold and as class 1 if it is greater than the threshold. Regression may also be described as finding that linear combination $\hat{y}$ of the predictor variables which has maximum correlation with the predictand y. Now the square of the correlation coefficient between y and $\hat{y}$ is an increasing function of the ordinary Student t-statistic comparing the two classes. And Fisher's linear discriminant analysis seeks to find that linear combination of the predictor variables which maximises just this squared correlation coefficient. It follows that the result obtained by regression will be the same as the result obtained by linear discriminant analysis. More formally, this result can be obtained as follows. We assume a regression model of the form $E(\mathbf{y}) = \mathbf{X}\boldsymbol{\beta}$, here the variables have been mean-centred by their overall mean vector $\bar{\mathbf{x}}$, leading to normal equations $(\mathbf{X}'\mathbf{X})\boldsymbol{\beta} = \mathbf{X}'\mathbf{y}$. Now, with the vector $\mathbf{y}$ scoring 0 for objects in class 0 and 1 for objects in class 1 we have (using the fact that $\mathbf{X}$ is mean-centred)

$$\mathbf{X}'\mathbf{y} = n_1(\bar{\mathbf{x}}_1 - \bar{\mathbf{x}}) = \frac{n_1 n_0}{n_1 + n_0}(\bar{\mathbf{x}}_1 - \bar{\mathbf{x}}_0)$$

Also, using the standard decomposition into within and between sums of squares and products, we have

$$\mathbf{X}'\mathbf{X} = (n_1 + n_0 - 2)\mathbf{S} + \frac{n_1 n_0}{n_1 + n_0}(\bar{\mathbf{x}}_1 - \bar{\mathbf{x}}_0)(\bar{\mathbf{x}}_1 - \bar{\mathbf{x}}_0)'$$

Putting these into the normal equations yields

$$\left[(n_1 + n_0 - 2)\mathbf{S} + \frac{n_1 n_0}{n_1 + n_0}(\bar{\mathbf{x}}_1 - \bar{\mathbf{x}}_0)(\bar{\mathbf{x}}_1 - \bar{\mathbf{x}}_0)'\right]\boldsymbol{\beta} = \frac{n_1 n_0}{n_1 + n_0}(\bar{\mathbf{x}}_1 - \bar{\mathbf{x}}_0) \quad (2.2)$$

Now, letting $\alpha = (\bar{\mathbf{x}}_1 - \bar{\mathbf{x}}_0)'\boldsymbol{\beta}$, we can write this as

$$\boldsymbol{\beta} = \left(\frac{1 - \alpha}{n_1 + n_0 - 2}\right)\frac{n_1 n_0}{n_1 + n_0}\mathbf{S}^{-1}(\bar{\mathbf{x}}_1 - \bar{\mathbf{x}}_0)$$

That is, $\boldsymbol{\beta}$ is proportional to $\mathbf{S}^{-1}(\bar{\mathbf{x}}_1 - \bar{\mathbf{x}}_0)$, which is what we wanted.

2.3 MORE THAN TWO CLASSES

When viewed from the perspective of assuming multivariate normal distributions, the extension of Fisher's method to more than two classes is straightforward. Another way of describing the classification rule derived from (2.1) is that we want to assign a new point $\mathbf{x}$ to the class j which has the largest value of

$$f(j|\mathbf{x}) \propto \frac{\pi_j}{(2\pi)^{d/2}|\Sigma|^{1/2}} \exp\left[-\frac{1}{2}(\mathbf{x} - \mu_j)'\Sigma^{-1}(\mathbf{x} - \mu_j)\right] .$$

Again log transforming, this is equivalent to assigning to the class which has the largest value of $\ln(\pi_j) + \mathbf{x}'\Sigma^{-1}\mu_j - \frac{1}{2}\mu_j'\Sigma^{-1}\mu_j$.

Alternatively, without making the multivariate normal assumption, we have the following extension. With two classes we found the direction which max-imised the ratio $[\mathbf{w}'(\bar{\mathbf{x}}_1 - \bar{\mathbf{x}}_0)]^2/\mathbf{w}'\mathbf{S}\mathbf{w}$. The numerator here is equal to $\mathbf{w}'(\bar{\mathbf{x}}_1 - \bar{\mathbf{x}}_0)(\bar{\mathbf{x}}_1 - \bar{\mathbf{x}}_0)'\mathbf{w}$ which is proportional to $\mathbf{w}'[\sum_{j=0}^{1}(\bar{\mathbf{x}}_j - \bar{\mathbf{x}})(\bar{\mathbf{x}}_j - \bar{\mathbf{x}})']\mathbf{w}$, where $\bar{\mathbf{x}}$ is the overall mean vector. The middle factor here permits immediate generalisation to an arbitrary number of classes, defining the *between-class matrix* $\mathbf{B}$ to be $1/C[\sum_{j=1}^{C}(\bar{\mathbf{x}}_j - \bar{\mathbf{x}})(\bar{\mathbf{x}}_j - \bar{\mathbf{x}})']$. We can then define the maximum separation between the C classes to be that which maximises the ratio $\lambda = \mathbf{w}'\mathbf{B}\mathbf{w}/\mathbf{w}'\mathbf{S}\mathbf{w}$, where $\mathbf{S}$ is again the sample estimate of the assumed common covariance matrix

$$\mathbf{S} = \frac{1}{\sum n_j - C}\left[\sum_{j=1}^{C}\sum_{i=1}^{n_j}(\mathbf{x}_{ij} - \bar{\mathbf{x}}_j)(\mathbf{x}_{ij} - \bar{\mathbf{x}}_j)'\right].$$

Informally, we are trying to find that direction $\mathbf{w}$ which maximises the separation between the sample means, standardised for the within-class relations between the variables.

Differentiating λ with respect to $\mathbf{w}$ and equating to zero yields $\mathbf{B}\mathbf{w} - \lambda\mathbf{S}\mathbf{w} = 0$. This two-sided eigenvalue equation has solutions for values of λ satisfying $|\mathbf{B} - \lambda\mathbf{S}| = 0$. That is, it has solutions for eigenvalues of $\mathbf{S}^{-1}\mathbf{B}$. The number of distinct eigenvalues is equal to the smaller of $C - 1$ and d. The eigenvector $\mathbf{w}_1$ corresponding to the largest of these eigenvalues (λ_1) is the direction which leads to maximum separation, as measured by λ, between the groups. This direction is the *first discriminant function* or *first canonical variate*. When there are only two groups, as in the previous section, this direction is the only discriminant function.

Subsequent eigenvectors $\mathbf{w}_k$ (subsequent discriminant functions or canonical variates) correspond to maxima of λ subject to constraints $\mathbf{w}_k'\mathbf{S}\mathbf{w}_h = 0$ for $h = 1$, $\ldots, k - 1$, so that $\mathbf{w}_k'\mathbf{X}$ and $\mathbf{w}_h'\mathbf{X}$ are uncorrelated for $h \neq k$. If we make the $\mathbf{w}_k$ unique by adding the constraint $\mathbf{w}_k'\mathbf{S}\mathbf{w}_k = 1$ then $\mathbf{W}\mathbf{S}\mathbf{W}' = \mathbf{I}$ where $\mathbf{w}_k'$ is the kth row of $\mathbf{W}$. Given that the canonical variates define the directions in which the groups are best separated, in the sense described above, it is natural to plot the data using these variates as axes. In particular, the two-dimensional space spanned by the first two variates is often used. When the objective of the analysis is interpretation of the canonical variates, instead of classification, the canonical variates are often adjusted to have zero overall sample mean and unit standard deviation.

When the objective is to build a model (for interpretation and understanding) rather than simply to construct an effective classification rule, one will be interested in eliminating superfluous variation by dropping those canonical variates which do not contribute to separation between the groups. That is, those sample canonical variates which can be attributed simply to random variation in the data, rather than representing a real spread of the group means in the relevant direction. This question is also pertinent when one wants to reduce the number of variables spanning the space containing the decision surface, as discussed in Section 9.1. Tests for the number of canonical variates which should be retained are based on the analogous problems of multivariate analysis of variance. Essentially they look at the sizes of the eigenvalues, eliminating those which are sufficiently small.

2.4   QUADRATIC DISCRIMINANT ANALYSIS

The decision surface in the preceding section is a linear function of the x_j. We can produce more flexible classification rules if we expand the space spanned by the measurements by including transformations of them. That is, by including extra, 'derived' variables, which are functions of the simple x_j and using a linear function of the original variables and these new ones. In particular, one obvious extension is to include second-degree terms: the squares and cross-products of the x_j. This changes the discriminant function from the form $\mathbf{x'w}$ to the form $\mathbf{x'w} + \mathbf{x'Wx}$, with $\mathbf{W}$ a matrix of weights for the $(d + 1)d/2$ second-degree terms. Although the resulting decision surface will remain linear in the space spanned by the new full set of variables, in the space spanned by the x_j alone it will be quadratic.

A common classification rule of this sort is obtained by relaxing the restriction above that the two covariance matrices be assumed equal. For example, adapting equation (2.1) in this way we have

$$\frac{f(1|\mathbf{x})}{f(0|\mathbf{x})} = \frac{\dfrac{\pi_1}{(2\pi)^{d/2}|\Sigma_1|^{1/2}} \exp\left[-\dfrac{1}{2}(\mathbf{x} - \mu_1)'\Sigma_1^{-1}(\mathbf{x} - \mu_1)\right]}{\dfrac{\pi_0}{(2\pi)^{d/2}|\Sigma_0|^{1/2}} \exp\left[-\dfrac{1}{2}(\mathbf{x} - \mu_0)'\Sigma_0^{-1}(\mathbf{x} - \mu_0)\right]}$$

leading to

$$\mathbf{x}'(\Sigma_1^{-1}\mu_1 - \Sigma_0^{-1}\mu_0) - \frac{1}{2}\mathbf{x}'(\Sigma_1^{-1} - \Sigma_0^{-1})\mathbf{x} + \ln(\pi_1/\pi_0) + \frac{1}{2}\ln(\Sigma_0/\Sigma_1)$$
$$- \frac{1}{2}\mu_1'\Sigma_1^{-1}\mu_1 + \frac{1}{2}\mu_0'\Sigma_0^{-1}\mu_0$$

Here the last four terms are independent of $\mathbf{x}$ and merely contribute to the threshold of the classification rule. The first term gives the coefficients of the linear contribution of the x_j and the second gives the quadratic contribution. This is the approach of *quadratic discriminant analysis* (QDA).

It is not the only way one might choose to define a quadratic rule—one could choose the $\mathbf{W}$ in some other way—just as one need not choose $\mathbf{w}$ in the simple linear rule by $\mathbf{w} \propto \mathbf{S}^{-1}(\bar{\mathbf{x}}_1 - \bar{\mathbf{x}}_0)$. (An example of another way to choose $\mathbf{w}$ is given in Section 3.3) The choice depends on the criterion one is seeking to optimise.

As in Section 2.3, if one focuses on the maximum of the posterior probabilities of class membership rather than simply the differences between them, the extension to more than two classes is straightforward.

Quadratic classification rules are clearly more flexible than linear rules. Indeed, by setting $\mathbf{W}$ to zero they reduce to linear rules. However, this does not mean they will necessarily outperform linear rules in practice. Quadratic rules involve d linear parameters, $(d + 1)d/2$ quadratic parameters and a threshold. In contrast, linear rules merely involve d linear parameters and a threshold. Unless d is very small, the move from linear to quadratic greatly enhances the possibilities of overfitting. In general a substantial sample size is needed in order to be able to capitalise on the increased power of quadratic rules.

2.5 STRUCTURED COVARIANCE MATRICES

Both linear and quadratic discriminant analysis estimate the decision surface via estimates of the covariance matrices. They are perfectly general in their use of these matrices, making no assumptions about the structures those matrices may have. Often, however, one has a priori ideas about the forms that the covariance matrices may take. Making use of such information leads to an effective reduction in the number of parameters which need to be estimated and hence can lead to improved classification accuracy (just as moving from quadratic to linear can increase the accuracy, despite the greater flexibility of the former).

2.5.1 Random effects models

An important situation where this arises is when the variables are measurements of the same thing taken at successive points on an underlying continuum. This continuum is often time, measurements being taken at successive times, and the objective is to identify (as early as possible) which of two or more groups a new profile belongs to. Another common situation arises when the underlying continuum is wavelength and the profiles correspond to spectra of samples. Then one often has far more measurements than design set elements so that the variable selection techniques of Sections 2.6 and 9.1 have to be used. In both of these examples one might reasonably expect successive measurements to be correlated. If one can postulate a realistic covariance structure, it can be used in the quadratic method outlined above.

One class of covariance structures, which might be appropriate in this case, arises from assuming a common underlying form for the profile of each object, but where these profiles depend on parameters which have been randomly selected from some distribution. This is a *random effects* model. Thus we might postulate that the design set profiles $\mathbf{x}_i$ have the form $\mathbf{x}_i = \mathbf{X}_i\boldsymbol{\beta} + \mathbf{Z}_i\mathbf{b}_i + \boldsymbol{\epsilon}_i$. Here $\mathbf{X}_i\boldsymbol{\beta}$ indicates which class the ith design set object belongs to and also models any fixed effects (for example, that the profiles in any one group have the same linear slope), and $\mathbf{Z}_i\mathbf{b}_i$ models, via the random effects $\mathbf{b}_i$, differences between the objects (for example, that the profiles of the objects have different intercepts, randomly selected from a normal distribution). The $\boldsymbol{\epsilon}_i$ represent additional random variation on top of that arising from the random effects. These are often taken to be independent from one measurement to another. The covariance matrix which results from this has the form $\Sigma_i = \mathbf{Z}_i\mathbf{B}\mathbf{Z}_i^T + \mathbf{E}_i$, where $\mathbf{B}$ is the covariance matrix of the $\mathbf{b}_i$. If the elements of $\boldsymbol{\epsilon}_i$ are taken to be independent then $\mathbf{E}_i$ is diagonal. Other common forms for $\mathbf{E}_i$ arise from:

(i) Autoregressive errors, based on the assumption that the jth element of $\boldsymbol{\epsilon}_i$ has the form $\varepsilon_j = \rho\varepsilon_{j-1} + u_j$, where the u_j are distributed as $N(0, \sigma_u^2)$. This yields an $\mathbf{E}_i$ matrix in which the jkth element has the form $E_{jk} = \sigma_u^2\rho^{|j-k|}/(1 - \rho^2)$.

(ii) Antedependence models in which the autoregressive process is extended to higher orders, the error variances may be unequal, and the 'time' intervals between measurements may be unequal.

2.5.2 Covariance selection models

The random effects models described above produce structured covariance matrices by virtue of the fact that the measurements are of the same kind and occur in some sort of sequence. An entirely different class of models is that of covariance selection models. These model the covariance matrix in terms of an underlying theoretical model of the relationships between the variables. For example, one might have ideas about the causal relationships between variables, and hence about the structure of the covariances between them.

If we assume that the measurement vector $\mathbf{x}$ for a particular class is multivariate normally distributed, then two variables are independent given the remaining variables if and only if the corresponding element of the inverse of the covariance matrix is zero. Beliefs about relationships between the variables—particularly about conditional independence relationships—can then be straightforwardly modelled by setting elements of the inverse of the covariance matrix to zero. This results in fewer parameters to be estimated.

2.5.3 Common principal components analysis

Linear discriminant analysis assumes that the covariance matrices for the various groups are the same. This is equivalent to requiring that the principal components of all the groups lie in the same direction and that the variances along these components are equal. We can relax this slightly by letting these variances differ from group to group. The result, called the *common principal components analysis* structure, is a model which lies between LDA and QDA in terms of its flexibility.

Yet another extension with flexibility lying between LDA and QDA arises by requiring the covariance matrices to be proportional. This obviously lies between the common principal components structure and LDA in terms of its flexibility.

2.6 REDUCING THE NUMBER OF VARIABLES

2.6.1 Principal components regression

Let $S = \sum_{k=1}^{d} e_k \mathbf{u}_k \mathbf{u}_k'$ be the spectral decomposition of the cross-product matrix $X'X$. In principal components regression (PCR) the matrix $S^{-1} = \sum_{k=1}^{d} e_k^{-1} \mathbf{u}_k \mathbf{u}_k'$ in the parameter estimator is replaced by by $\tilde{S}^{-1} = \sum_{k=1}^{m} e_k^{-1} \mathbf{u}_k \mathbf{u}_k'$, where the summation is over the 'first' m components. 'First' here may refer to those with the largest eigenvalues or it may refer to those m which have the highest correlation with the dependent variable. The latter is arguably the more sensible approach since there is no a priori reason why components with large eigenvalues should be strongly related to the dependent variable.

In effect, this approach reduces the dimensionality of the predictor space so that high correlations and multicollinearity problems are alleviated.

2.6.2 Partial least squares regression

Linear regression analysis and linear discriminant analysis find the single linear combination of the predictor variables which has maximum correlation with the response or class variable. This single linear combination will use all of the predictor variables or will use a subset (perhaps selected by one of the methods described in Section 9.1, such as stepwise selection). In contrast, like principal components regression, partial least squares regression reduces the number of variables on which the ordinary least squares regression is carried out. Also like PCR, this is done by 'feature extraction' rather than variable selection; that is, it produces 'derived variables' from the raw measured ones, rather than simply selecting subsets (see Section 9.1). The first of these derived variables is the linear combination of predictor variables which has maximum *covariance* (rather than correlation) with the dependent variable. Subsequent derived variables are found in the same way, subject to additional constraints of being uncorrelated with previous ones. The result is a reduced set of derived variables which can be used to predict the dependent variable by ordinary regression. Of course, once the process is finished and the final ordinary least squares regression on the partial least squares components performed, the result can be re-expressed as a linear combination of the raw variables.

The derived variables extracted by this process are called partial least squares components and can be thought of as latent factors. A factor analysis of the raw data matrix $\mathbf{X}$ can be written as

$$\mathbf{X} = \mathbf{F}\Lambda + \mathbf{U} = \{\mathbf{f}_1, \ldots, \mathbf{f}_q\} \begin{pmatrix} \mathbf{l}'_1 \\ \vdots \\ \mathbf{l}'_q \end{pmatrix} + \mathbf{U} = \sum_k \mathbf{f}_k \mathbf{l}'_k + \mathbf{U}$$

where $\mathbf{f}_k$ is the kth column of $\mathbf{F}$ and is the kth factor, and $\mathbf{l}'_k$ is the kth row of Λ, the loading matrix. Then $\mathbf{y}$ can be expressed in terms of these factors as

$$\mathbf{y} = \{\mathbf{f}_1, \ldots, \mathbf{f}_q\} \begin{pmatrix} \beta_1 \\ \vdots \\ \beta_q \end{pmatrix} + \mathbf{e}$$

where the β_k are scalars.

Partial least squares and ordinary least squares methods have been combined into a single unified approach called continuum regression by defining a general criterion (covariance)^2 (variance)$^{a/(1-a)-1}$. When $a = 0$, this yields ordinary least squares and when $a = 1/2$ it yields partial least squares (and when $a \approx 1$ it effectively yields the principal components).

2.6.3 SIMCA

SIMCA is possibly the most widely used member of the class of classification rules known as *subspace methods*, which are based on the idea of representing each class, separately, by its own subspace. SIMCA stands for Soft Independent Modelling of Class Analogy, perhaps not the most expressive of descriptions. In the two-class case, simple linear methods use a single set of variables (for example, all the measured variables or the first few principal components) to produce a single discriminant function which, by comparison with a threshold, serves to define the decision surface. The single discriminant function is estimated essentially by contrasting the two design set classes (the precise details depend on the method adopted). In contrast, SIMCA selects *two* separate subsets of derived variables, one for each class, defined as the first few principal components for each class separately. Using these to define two subspaces, a new object is compared with each subspace to assess its similarity to each set. In particular, for class 0 the ratio of the distance of the new measurement vector from the class 0 subspace to the root mean square distance of class 0 design set objects from the class 0 subspace is calculated. This gives a distance measure standardised for the variation of class 0 design set objects. (In fact it gives a

Mahalanobis distance in which the eigenvalues associated with the subspace are all set to be infinite and those associated with the complementary space are taken to be a class-dependent constant.) A similar calculation is undertaken for class 1 and the two standardised distances are compared. Cross-validation has been proposed as a way of choosing the number of components in each set.

By explicitly computing a measure of distance of the point to be classified from each class separately, the SIMCA method allows the natural introduction of atypicality indices. And generalisation to more than two classes is trivial, again by virtue of the fact that the distance of the point to be classified is estimated for each class separately.

2.6.4 DASCO

DASCO, Discriminant Analysis with Shrunken COvariances, competes with the acronym SIMCA for being uninformative, but the technique does overcome SIMCA's two weaknesses. SIMCA ignores information about differences between the groups in the subspaces spanned by the sets of first few principal components (since it effectively takes the corresponding within group eigenvalues to be infinite). And because it assesses the similarity of a new object to the design set objects in each class using a Mahalanobis distance, SIMCA effectively ignores the factors $|\Sigma_j|^{-1/2}$ for each class. These would be required to get accurate probability estimates under the assumption of multivariate normality. DASCO overcomes the first of these weaknesses by including, for each class, the eigenvalues from the first few principal components. It also slightly modifies the other eigenvalues, but again takes them to be constant. With this choice, new estimates of the covariance matrices are produced which can be used to compute distances of the new object from each class. The second weakness is overcome by using probability estimates assuming multivariate normal distributions using these covariance matrices, instead of merely the Mahalanobis distance.

2.7 REGULARISATION

Regularisation methods (the term comes from regularisation methods in approximation theory) shrink a highly parametrised model towards some simpler, less highly parametrised model. The degree of shrinkage is chosen by fixing a *regularisation parameter*. Shrinking from a highly parametrised model to a simpler model introduces the risk of biasing the estimator. However, it is well known that shrinkage can reduce loss functions such as mean square error, but it is not clear, a priori (and theoretical analysis is intractable) that it will improve a criterion such as error rate. Simulation and practical experience, however, suggests that it often does. In this section we examine *ridge regression*, an adaptation of ordinary least squares regression to the case when the predictor variables are multicollinear, and *regularised discriminant analysis*, a method specially designed for discriminant analysis. Both methods focus on the large number of parameters required in an unconstrained covariance matrix.

2.7.1 Ridge methods

Unlike OLS estimates, ridge estimates are biased in a deliberate attempt to effect a reduction in variance so that, overall, the mean squared error is reduced. Ridge regression does not predict the design data as accurately as OLS in that the sum of squared residuals is larger, but the method can produce greater accuracy in predicting new values, which is what really interests us.

Ridge regression modifies the standard OLS solution by altering the cross-product matrix in the least squares solution so that $\hat{\mathbf{w}} = (\mathbf{X}'\mathbf{X} + k\mathbf{I})^{-1}\mathbf{X}'\mathbf{y}$. Here $k > 0$ is a parameter which causes the parameter estimates to shrink towards zero as k grows. When $k = 0$ we obviously have the OLS solution. $\mathbf{X}'\mathbf{X} + k\mathbf{I}$ is effectively a weighted sum of the usual $\mathbf{X}'\mathbf{X}$ and $\mathbf{I}$. Another way of describing the ridge solution is as the estimate which minimises $\|\mathbf{y} - \mathbf{Xw}\|^2 + k\|\mathbf{w}\|^2$, where $\mathbf{y}$ and $\mathbf{X}$ have been mean centred. In practice, obtaining ridge estimates is a rather subjective process. Plots are made of (i) the estimated weights w_k against k and (ii) the residual sum of squares against k. That value of k is chosen which leads to stable coefficient plots, normally called *traces*, without penalising the sum of squares too much.

The result of this modification on linear discriminant analysis can be easily seen using the equivalence of LDA and OLS, described in Section 2.2. If we add $k\mathbf{I}$ to the $\mathbf{X}'\mathbf{X}$ in the normal equations in (2.2), a little algebra shows that the replacement matrix for $\mathbf{S}$ in the linear discriminant function has the form $(\mathbf{S} + k^*\mathbf{I})$, with $k^* = k/(n_1 + n_0 - 2)$.

When the rank of $\mathbf{S}$ is $r < d$ an alternative is to replace $\mathbf{S}$ by $\{\mathbf{I} - \mathbf{L}_1\Delta[(d_k - \beta)/(d_k + \alpha)]\mathbf{L}_1'\}$, where $\Delta(s_k)$ is a diagonal matrix with kth term s_k, $\mathbf{L}_1$ is the matrix with columns the first r eigenvectors of $\mathbf{S}$, and d_k are the corresponding eigenvalues. The parameters α and β are determined by cross-validation to minimise error rate.

2.7.2 Regularised discriminant analysis

The covariance matrix of the jth class can be represented in terms of its spectral decomposition by $\Sigma_j = \sum_{k=1}^{d} e_{jk}\mathbf{v}_{jk}\mathbf{v}_{jk}'$ where e_{jk} is the kth eigenvalue for the jth class (ordered by decreasing size) and $\mathbf{v}_{jk}$ is the corresponding eigenvector. Using this decomposition, we have $\Sigma_j^{-1} = \sum_{k=1}^{d} \mathbf{v}_{jk}\mathbf{v}_{jk}'/e_{jk}$. Given the role that Σ_j^{-1} plays in discriminant analysis, this means that the directions corresponding to the smallest eigenvectors will have a large impact on the direction of the resulting discriminant function. Now, when the maximum likelihood estimate is used for the covariance matrix, the estimates of the eigenvalues are biased. In particular, the estimates of the smallest eigenvalues are biased low. The overall consequence is to unduly emphasise the directions corresponding to the smallest eigenvalues, so leading to most of the variance in the discriminant scores. *Regularised discriminant analysis* seeks to overcome this by two steps.

First, a regularisation of quadratic discriminant analysis is made by replacing $\hat{\Sigma}_j = \mathbf{S}_j / n_j$ (where $\mathbf{S}_j$ is the class j cross-product matrix and n_j is the number of design set points in class j) by

$$\Sigma_j^*(\lambda) = \frac{(1 - \lambda)\mathbf{S}_j + \lambda\mathbf{S}}{(1 - \lambda)n_j + \lambda n}$$

where $\mathbf{S} = \sum_j \mathbf{S}_j$ and $n = \sum_j n_j$. That is, it shrinks the crude maximum likelihood estimate of each class's covariance matrix towards the common average value. The extent of shrinkage can be changed by varying the regularisation parameter λ between 0 and 1. A second stage of shrinkage, replacing the above by $(1 - \gamma)\Sigma_j^*(\lambda) + \gamma d^{-1}\text{tr}(\Sigma_j^*(\lambda))\mathbf{I}$ shrinks things towards the identity matrix $\mathbf{I}$. The two parameters λ and γ can be estimated by cross-validation using the misclassification rate as the performance criterion.

2.8   SHORTEST LEAST SQUARES

A common way of overcoming the singularity problems arising from a small number of design set elements relative to a large number of measured variables is some kind of variable selection or extraction prior to application of a 'standard' method. This is not the only possible approach. In regression terms, singularity of the covariance matrix of 'predictor' variables means that an infinite number of sets of weights will satisfy $\hat{y} = \mathbf{w}'\mathbf{x}$. A unique set can be defined by imposing extra constraints on $\mathbf{w}$. For example, in *shortest least squares* one chooses that $\mathbf{w}$ for which $\mathbf{w}'\mathbf{w}$ is a minimum. This yields $\hat{\mathbf{w}} = (\mathbf{X}'\mathbf{X})^+\mathbf{X}'\mathbf{y}$, where $(\mathbf{X}'\mathbf{X})^+$ is the Moore–Penrose generalised inverse of $\mathbf{X}'\mathbf{X}$.

2.9   FURTHER COMMENTS

The methods described in this chapter have at their root the estimation of the class conditional probability distributions $f(\mathbf{x}|j)$ or the relationships (e.g. distances) between such distributions. From these, the posterior probabilities $f(j|\mathbf{x})$ are estimated using Bayes' theorem. In particular, we showed that if multivariate normal distributions with equal covariance matrices are assumed then the optimal decision surface is linear. A linear decision surface will also result if the conditional distributions are assumed to be of the more general form $N(\mu_j, \Sigma)g(\mathbf{x})$, where g is some common function. However, in this case simple LDA may well not to lead to a good decision surface; the covariance matrices of the distributions for each group may be quite different, depending upon g. In contrast, methods which are based directly on estimates of the $f(j|\mathbf{x})$ might be expected to do well. Examples of such methods are logistic regression and the error-correcting estimation methods used in neural networks.

Since, as we have shown above, the quadratic method is optimal when the class conditional distributions are multivariate normal, a possible general approach is to transform the data to normality before estimating the decision surface. In general, finding multivariate transformations to multivariate normality is not tractable, but one could transform the marginals to univariate normality (though this does not guarantee multivariate normality). An obvious approach is to use the Box and Cox family of transformations, given by

$$
x^\lambda = \begin{cases} \dfrac{x^\lambda - 1}{\lambda} & \lambda \neq 0 \\ \log(x) & \lambda = 0 \end{cases}
$$

The likelihood of $\mathbf{x}$ for the jth class is then

$$
\frac{\pi_j}{(2\pi)^{d/2}|\Sigma|^{1/2}} \exp\left[-\frac{1}{2} (\mathbf{x}^{(\Lambda)} - \mu_j)' \Sigma_j^{-1} (\mathbf{x}^{(\Lambda)} - \mu_j) \right] \prod_{k=1}^{d} x_k^{\lambda_k - 1}
$$

where $\mathbf{x}^{(\Lambda)}$ represents the vector of transformed xs. As before, we can then take logs and classify according to the maximum of these expressions. This differs from the situation immediately above in that the $\prod_{k=1}^{d} x_k^{\lambda_k - 1}$ terms will normally differ from class to class.

When $\mathbf{X}'\mathbf{X}$ is nonsingular, ordinary least squares regression (and therefore LDA) is invariant to rescaling (or more general linear transformations) of the predictor variables. However, regularisation methods involve using some extra parameters, not dependent on the relationship to the $\mathbf{y}$ variable. As a consequence, they are not invariant to rescaling. (Except in certain special cases. If PLS and PCR use as many factors as possible, they become identical to OLS.)

This book is about classification rules—rules which may be used to classify future objects to a class. LDA may be considered to identify that direction in measurement space in which the classes are most widely separated and then to define a decision surface orthogonal to this direction. This separating direction in some sense characterises the relationship between the classes. Indeed, another common use of discriminant analysis is as an aid to understanding *how* classes differ. The sizes of the coefficients of a linear discriminant function serve to identify which variables (and combinations of variables) are important in distinguishing between classes. Moreover, having identified the direction in which the classes most differ, one might legitimately ask how much they differ—or if the samples are 'significantly different'. That is, one might be interested in making inferences about the populations from which the samples have been drawn. As we remarked above, such questions are answered using the techniques of multivariate analysis of variance. Multivariate analysis of variance finds exactly the same linear combinations of the variables as LDA, but then uses them to answer questions about how the groups differ and the extent and significance of the differences.

Several authors have considered robustness of linear discriminant analysis to non-multivariate normal measurement distributions. Section 5.6 of McLachlan (1992) reviews such work.

In many discriminant analysis situations the classes are qualitatively distinct. Examples are male versus female, dead versus living, and so on. In other situations, however, the classes are defined by imposition of a cutoff point on some underlying continuum. Examples are old versus young and the classification into mild or severe, common in medical situations. Sometimes the score on the latent continuum is never explicitly produced and, indeed, it might be the case that explicitly measuring this score is very difficult or even impossible so that only the classification is observed. But the point is that, in such cases, the underlying continuum, on which a threshold has been imposed, is assumed to exist. Given such an underlying continuum which defines the classes by a threshold, a reasonable assumption would be that the joint distribution of the measurements and the continuum is multivariate normal. If this is assumed, it follows that the marginal multivariate distribution of the measurements for each of the classes separately *cannot* in general be ellipsoidal and cannot, for example, be multivariate normal. This is a common situation, so it is pertinent to ask, How would linear discriminant analysis be expected to perform in these circumstances? In particular we may ask, How does the decision surface produced by this common method compare with the optimal decision surface? This question is discussed in Section 9.4.

2.10 FURTHER READING

Fisher (1936) described the fundamental approach to linear discriminant analysis outlined above. Although, as this chapter has illustrated, the basic form has been extended in many ways, even the simple form continues to be widely used and to be of great value. It is available in all the major software packages. The performance of Fisher's LDA with nonnormal classes has been investigated by several authors, for example Gilbert (1968), Moore (1973) and Krzanowski (1977).

Krzanowski (1989) and Ringrose and Krzanowski (1991) have developed methods for producing confidence ellipsoids for the points representing group means in two-dimensional plots from a discriminant analysis.

A test for irrelevant canonical variates, in terms of insignificant contribution to the separation between groups, is described in Bartlett (1947, 1951).

Random effects models are described in Crowder and Hand (1990), Longford (1993) and Hand and Crowder (1996). In particular, Chapter 6 of Hand and Crowder (1996) shows how various covariance matrices can arise. Work on covariance selection models is described in Dempster (1972), Speed and Kiiveri

(1986), Whittaker (1990) and Wermuth (1980, 1991). Common principal components analysis models are described in Flury (1988) and, in particular, the application to discriminant analysis in Section 2.4 of Flury (1995).

Partial least squares was originally developed by H.Wold and his colleagues (H.Wold, 1966, 1985) and approached from an algorithmic perspective. The description in terms of regularisation by maximising covariance is due to Frank (1987), Helland (1988) and Stone and Brooks (1990). Stone and Brooks (1990) also introduced the idea of continuum regresssion to unify OLS and PLS.

Stone and Jonathan (1993, 1994) provide an excellent review of applications of statistics in quantitative structure–activity relationship studies, which are prime examples of the sorts of problems described in this chapter. Parts of this chapter draw heavily on their work.

The SIMCA method was originally described by S.Wold (1976), although the idea of describing each class using separate subspaces had been suggested earlier (Watanabe, 1965, 1970). The method has been enthusiastically adopted by the chemometric community; see, for example, Kowalski and Wold (1982). DASCO was introduced by Frank and Friedman (1989) in order to overcome the two weaknesses of the SIMCA method mentioned above. After a series of simulation studies and applications to real data they concluded, 'RDA and DASCO provide equal or better prediction than LDA or QDA, and much better than that of SIMCA. This is especially true as the observation/variable ratio decreases ... RDA has the advantage over DASCO that, in addition, it can shrink the class covariance matrices toward the pooled covariance matrix. This feature gives RDA an advantage when the class covariance matrics tend to be similar. RDA is also much more rapidly computable than either SIMCA or DASCO.' An introduction to subspace methods in general, including SIMCA, is given by Oja (1983).

Ridge methods of discriminant analysis have been explored by several authors, for example Campbell (1980) and Rodriguez (1988). Peck et al. (1988) compared some of these approaches. The method of ridge discriminant analysis described above for the case of singular cross-products matrices was described by Krzanowski et al. (1995) in an interesting paper in which they compare several approaches to this problem. Their conclusion was that no single method dominated the others and they recommended the accumulation of practical experience to build up a database from which recommendations could be made.

Shrinkage approaches to linear discriminant analysis were explored by Peck and van Ness (1982) and 'regularised discriminant analysis', as described above, was introduced by Friedman (1989) who pointed out that the case $(\lambda, \gamma) = (0, 0)$ corresponds to QDA, (λ, γ), $\lambda = (1, 0)$ corresponds to LDA, and $(\lambda, \gamma) = (1, 1)$ corresponds to the nearest means classifier (assigning an object to the class with the nearest design set mean, using Euclidean distance). Less extreme models are achieved by varying the parameters. Aeberhard et al.

(1993) claim to improve the performance of the method by using smoothed estimates of error rate as the criterion in estimating the regularisation parameters.

The idea of transforming the marginal xs to normality is described by Velilla and Barrio (1994). Multivariate analysis of variance is described in Hand and Taylor (1987).

CHAPTER 3
Nonlinear Methods

3.1 INTRODUCTION

In recent years, neural networks have attracted colossal interest. Some protagonists have promoted them as a universal answer. Since one of their most important application domains is classification problems, it is especially relevant for us to see why they have attracted so much interest.

Firstly, the name could hardly have been better designed to attract interest (analogous to the term 'electronic brains' in the early days of computers). Secondly, they apparently have the ability to 'learn' how to perform some task, such as classification, without being given specific instructions. (Of course, one may ask what is the difference between 'learning' and the less glamorous sounding 'recursive parameter estimation'.) Thirdly, early studies suggested that they performed such tasks well. Fourthly, they permit an attractive graphical representation. Finally, and perhaps most important of all, they were applied to problems which statisticians regarded as difficult, possibly due to 'messy' data.

The fact is, however, that neural networks are but one family from a class of statistical models which might be described as occupying an intermediate position on a continuum of simplicity/complexity. With progress in computer technology and the consequent growing incidence of large data sets, it has become more important that this gap should be filled. We can define the two ends of this continuum as follows.

At the simple end lie the linear models described in Chapter 2. They sometimes make Draconian assumptions about the decision surface. To the extent that these assumptions are not satisfied, the classifiers will be poor. One way to overcome any inadequacies of linear methods applied directly to the measured variables is to include transformations of those variables in the linear combination. Thus, for example, quadratic decision surfaces are linear in the raw measured variables, their squares and their two factor products. Nonetheless, the forms of the decision surfaces taken by such methods are fairly restricted—the models do not permit small irregularities in the decision surface, for example.

At the other extreme are nonparametric classification rules based on the idea of finding local approximations to the class conditional probabilities $f(j|\mathbf{x})$, as

discussed in Chapter 5. Essentially these work by averaging the classes of those
design set points nearest to the point x_0 at which a classification is required.
In order to do this without the risk of introducing large bias, we require that
the points which are averaged are near to x_0. If they are far from x_0 then it
is likely that the conditional probabilities $f(j|x)$ will be very different
from $f(j|x_0)$, so that the average could also be very different from
$f(j|x_0)$. Unfortunately, as the dimensionality increases, 'near' very rapidly
becomes 'far'—a consequence of the curse of dimensionality. An exponential
increase in the number of data points is needed to compensate for this.
Nonparametric methods are very flexible but can pay for this by requiring large
sample sizes. (Of course, in principle, one could make a linear model have
equivalent flexibility by adding arbitrarily many polynomial terms to the model.
But this would have the cost of giving flexibility where it is not required, which
seems extravagant.)

Neural networks thus occupy an intermediate position on the continuum of
model simplicity/complexity. In a very loose sense they have filled a gap between
the inflexible and the too flexible. We describe neural networks in Section 3.3;
later sections describe other models aimed at filling this same flexibility gap and
mostly developed within the statistical community. But to set the scene, Section
3.2 describes the method of logistic regression. Logistic regression has been
around for several decades. It is best regarded as a more natural model than the
simple linear model for the conditional probability that an object with measurement vector x will belong to each of two classes. However, it is also possible to
regard it as a basic building block from which the flexible models of neural
networks are typically constructed.

3.2 LOGISTIC REGRESSION

Consider the case of two classes. We could model $f(1|x)$ as a linear function
of the components of x, $\hat{f}(1|x) = \beta'x$ but this would have the obvious weak-
ness that the estimate would not lie between 0 and 1. This may not matter
for classification purposes; if classifications are to be induced by comparison
with a threshold, or by rank ordering the estimated probabilities, the restriction
to the interval $[0,1]$ is unnecessary. On the other hand, it is an unappealing
solution; if the model does not faithfully represent this particular property of
the underlying situation, then what other properties does it fail to faithfully
represent? Moreover, if the parameters of the model are to be estimated by
minimising some goodness of fit criterion, measuring the discrepancy between
the data and the theoretical distributions arising from the model, then questions
are raised about the whole process if the model is obviously wrong from the
start.

We can sidestep this difficulty by transforming the linear combination so that
it always lies between 0 and 1. Put the other way, we can transform the prob-

ability $f(1|\mathbf{x})$ so that the interval $[0,1]$ is mapped to the range $(-\infty, +\infty)$. Several such transformations have been widely used, including

- *Probit*: the linear combination estimates $\Phi^{-1}(f(1|\mathbf{x}))$, where Φ is the cumulative normal distribution, so that $f(1|\mathbf{x}) = \Phi(\beta'\mathbf{x})$.
- *complementary log-log*: the linear combination estimates $\log[-\log(1 - f(1|\mathbf{x}))]$ with inverse $f(1|\mathbf{x}) = 1 - \exp(-\exp(\beta'\mathbf{x}))$.
- *logistic*: the linear combination estimates $\log[f(1|\mathbf{x})/(1 - f(1|\mathbf{x}))]$ with inverse $f(1|\mathbf{x}) = \exp(\beta'\mathbf{x})/[1 + \exp(\beta'\mathbf{x})]$.

Although each transformation arises naturally in some circumstances, the logistic transform is nowadays the approach used most generally and most widely. Unlike the complementary log-log, the logistic transform is symmetric; it also has a natural interpretation in terms of log odds ratios and arises naturally in some important special cases, as we shall see below.

The parameters in the logistic model can be estimated (iteratively) by maximising the likelihood function $\prod_{i=1}^{n} \hat{f}(1|\mathbf{x}_i)^{c_i}(1 - \hat{f}(1|\mathbf{x}_i))^{1-c_i}$, where $c_i \, (= 0, \, 1)$ is the true class of the ith object and $\hat{f}(1|\mathbf{x}_i)$ is its estimated probability of belonging to class 1, a function of the parameters β.

Apart from the fact that estimation has to be iterative, the nonlinearity in the above model has other implications. One is that notions of orthogonality, so important with balanced linear models (for example, permitting unambigous partitioning of sums of squares in analysis of variance into components uniquely attributable to each of the factors), cease to have meaning. The implication is that the coefficients must always be interpreted in the context of the other coefficients in the model.

Although the model here is based on an underlying linear combination of the raw variables, one will often want to increase the flexibility of the model by also using higher powers and cross-products of the raw variables—in just the same way that the flexibility of Fisher's linear discriminant analysis can be extended.

One reason for the widespread adoption of logistic regression in the context of allocation problems can be seen by considering the special case of two multivariate normal populations, with equal covariance matrices. In this case the logistic transformation above becomes

$$
\log\left[\frac{f(1|\mathbf{x})}{1 - f(1|\mathbf{x})}\right] = \log\left[\frac{\dfrac{\pi_1}{(2\pi)^{d/2}|\Sigma|^{1/2}}\exp\left[-\dfrac{1}{2}(\mathbf{x} - \mu_1)'\Sigma^{-1}(\mathbf{x} - \mu_1)\right]}{\dfrac{\pi_0}{(2\pi)^{d/2}|\Sigma|^{1/2}}\exp\left[-\dfrac{1}{2}(\mathbf{x} - \mu_0)'\Sigma^{-1}(\mathbf{x} - \mu_0)\right]}\right]
$$

$$
= \log\frac{\pi_1}{\pi_0} - \frac{1}{2}(\mathbf{x} - \mu_1)'\Sigma^{-1}(\mathbf{x} - \mu_1) + \frac{1}{2}(\mathbf{x} - \mu_0)'\Sigma^{-1}(\mathbf{x} - \mu_0)
$$

$$
= \mathbf{x}'\Sigma^{-1}(\mu_1 - \mu_0) + \ln(\pi_1/\pi_0) - \frac{1}{2}\mu_1'\Sigma^{-1}\mu_1 + \frac{1}{2}\mu_0'\Sigma^{-1}\mu_0
$$

which is exactly the same as the linear discriminant function.

In view of the above result, one might ask what is the difference between the two approaches, when applied to the multivariate normal case. The answer is that linear discriminant analysis uses more information. The logistic method directly estimates the (logistic transform of the) conditional probabilities $f(1|\mathbf{x})$, whereas the linear discriminant approach derives these probabilties indirectly via estimates of the multivariate normal class conditional distributions. Put another way, the logistic method directly estimates the linear weights $\beta = \Sigma^{-1}(\mu_1 - \mu_0)$ whereas the linear discriminant approach derives them as appropriate combinations of estimates of Σ and the μ_i. In practice the results are usually very similar.

As a second special case, consider the situation when the predictors are independent binary variables. Letting p_k be the probability that the kth variable x_k takes the value 1 for objects in class 1 and r_k the corresponding probability for objects in class 0, we have

$$\log\left[\frac{f(1|\mathbf{x})}{1 - f(1|\mathbf{x})}\right] = \log\left[\frac{\prod_{k=1}^{d} p_k^{x_k}(1 - p_k)^{1-x_k}}{\prod_{k=1}^{d} r_k^{x_k}(1 - r_k)^{1-x_k}}\right]$$

$$= \sum\left\{x_k\log\left(\frac{p_k}{r_k}\right) + (1 - x_k)\log\left(\frac{1 - p_k}{1 - r_k}\right)\right\}$$

$$= \sum x_k\log\left(\frac{p_k}{r_k}\cdot\frac{1 - r_k}{1 - p_k}\right) + \sum\log\left(\frac{1 - p_k}{1 - r_k}\right)$$

so that the optimal decision surface is linear, with the coefficients of the x_k being log odds ratios.

The discussion above has been restricted to two classes. Extension to more than two classes is straightforward and can be done in several ways, depending on whether or not the classes have a natural order. If they do not have a natural order, it is possible to pick one of the classes (class 1, say) as a baseline and fit separate linear models to $\log f(j|\mathbf{x})/f(1|\mathbf{x})$. Sometimes this is called a *multiple logistic model*. If the classes have a natural order, one can model $\log f(j|\mathbf{x})/\sum_{s=1}^{j-1} f(s|\mathbf{x})$.

3.3 NEURAL NETWORKS

3.3.1 Introduction

The original motivation for research in neural networks came from an interest in understanding and modelling the way organic brains function. This motivation continues, and has led to its own flavour of work. However, in terms of research effort and resource expenditure it has been far exceeded by work in a slightly different direction: the construction of networks to tackle engineering, technological and scientific problems, whether or not they mimic brain architectures. In particular, a vast amount of work has been done on building networks to estimate mathematical functions. In this book we are concerned with that segment of the work (a very large segment) in which the function to be estimated is a classification function.

Experiment shows that brain cells perform their elementary operations in times of the order of milliseconds. This is very slow compared with modern computers, the elements of which perform their operations in times of the order of nanoseconds. And yet the brain can process data vastly faster than even the fastest of modern computers. Some clarification is needed here: the reader might justifiably object that a computer (even the slowest) can add a column of figures much more rapidly than can a human, and can carry out, in the twinkling of an eye, other arithmetical and symbol manipulation activities which would take humans hours or days. This is true, but all these problems are well defined, involving only a small alphabet of symbols, for which the manipulations to be undertaken are clearly specified, and which do not involve complex search procedures. In contrast, activities such as face recognition, speech understanding and playing tennis involve interactions with the uncertainties of the real world and a huge amount of incoming stimuli which have to be sorted and selected (not merely a column of figures). We do these sorts of things effortlessly (though some with more skill than others, as in playing tennis) but programming computers to do them has posed a research challenge which has lasted for decades.

If the basic elements of brains function so slowly, how is it that brains can perform these mammoth data-processing operations so quickly? The answer lies in the number of individual elements in the brain, the extraordinarily rich complexity of connections between them, and the fact that they tend to work simultaneously (in parallel), not sequentially (one after the other) as in digital computers. For these reasons research in this area is sometimes described by the nouns *parallelism* or *connectionism*. Practical implementations of such systems, aimed at tackling real problems, have the form of an array of simple processors connected by 'weighted' links. Input to a particular processor comes via its links with neighbouring processors. It carries out some (simple) procedure on its inputs (such as adding them up and transforming the result) and sends the results along links to neighbouring processors. Some architectures involve directed links—inputs and outputs come from, and go to, different processors—whereas others involve undirected links. Directed link architectures have been the most widely developed for classification problems: the input and output sets of a given processor are disjoint sets.

The description above, in terms of processors connected by weighted links, suggests a natural representation as a *graph* or *network* structure, with the processors being *nodes* of the graph and the links being *edges* of the graph. This leads to convenient graphical (diagrammatic) representations. These representations are similar to structures such as the covariance selection models described in Section 2.5, which can also be portrayed as mathematical graphs. We describe neural networks in the following section, beginning with the *Rosenblatt perceptron*, the earliest system of its kind.

3.3.2 The Rosenblatt perceptron

Figure 3.1 shows the basic structure of the Rosenblatt perceptron, the simplest kind of neural network. It has a single node which takes many inputs, processes them and delivers one output. The inputs, denoted x_k, are multiplied (weighted) by the w_k associated with the links, then simply summed by the processor. A classification is achieved by comparing the weighted sum of inputs with a threshold. In mathematical terms, the model has the form: if $\sum_{k=1}^{d} w_k x_k > t$ then classify into class 0 (say), otherwise into class 1. Here t is the threshold. We see that this simple network in fact produces a linear classifier. In statistical terms the weights w_k are parameters. We aim to choose those parameter values which yield the best classification performance. Various terminologies arise in this context, depending upon the background of the earlier investigators, but meaning the same thing. Statisticians will speak of estimating the parameters (almost with an implication that there is a 'true' value to be estimated). In contrast, those from a computer science or machine learning background may speak of training the system or of the system learning. Nowadays the various research communities have begun to recognise the contribution that the others have made and the terms may be used interchangeably.

Sometimes it is convenient to add an extra input x_0, with value fixed at 1 for all cases. This is then weighted by $-t$ so that a classification is obtained by comparing $\sum_{k=0}^{d} w_k x_k$ with 0. This makes no essential difference to the ideas. The complete vector, including the constant value 1, is sometimes called the

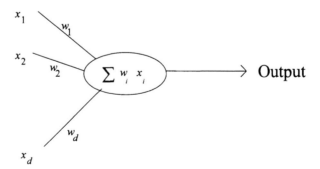

Figure 3.1   The Rosenblatt perceptron. elements of the input vector **x** are weighted by weights in vector **w** and summed to yield an output.

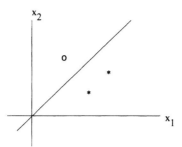

Figure 3.2 A line separating samples from two classes in the space spanned by the measurements x_1 and x_2.

augmented observation vector.

Diagrammatically we can represent a simple two-dimensional problem as in Figure 3.2. The axes are the variables, the stars and circles represent the positions of the objects in the space of the measured variables, and the line shows the decision surface. If the decision surface separates the points from the two classes (as it does in Figure 3.2), then $\mathbf{w}'\mathbf{x}$ is positive for the points from one class and negative for those from the other. Because the space in this representation is spanned by the measured variables, it is called the *measurement space representation*.

There is also a dual representation which often proves enlightening. The expression $\sum_{k=0}^{d} w_k x_k = \mathbf{w}'\mathbf{x}$ above is symmetric in its use of the x_k and the w_k so we could just as easily plot the w_k as axes. In the space spanned by these axes a possible decision surface is represented as a point. Objects (represented as points in the measurement space) become lines (more generally, hyperplanes) in this representation. This dual representation is termed the *weight space representation*. The weight space representation corresponding to Figure 3.2 is given in Figure 3.3, with the small line segments indicating the positive orientation of the

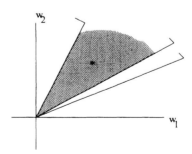

Figure 3.3 The corresponding representation in the weight space; points from Figure 3.2 become lines and vice versa.

lines. The star representing the decision surface from Figure 3.2 is on the positive side of the two lines corresponding to the star points and the negative side of the line corresponding to the circle point, corresponding again to perfect separability of the two design samples. Indeed, any point in the shaded region satisfies this condition, so the region is called the *separability region*.

We now need to estimate the weights from a design set of cases with known measurement vectors and class memberships. Early work on perceptrons within the computer science community placed great emphasis on perfectly separable classes. That is, it was assumed that the two classes could be perfectly separated by a linear surface in the measurement space, and the problem was seen as identifying such a perfectly separating surface. To a large extent this emphasis has continued and is reflected in much work on computational learning theory. In contrast, statistical work has always considered problems in which perfect separation may not be possible.

Generalisability issues aside, we shall seek a decision surface which separates the two design set classes as well as possible. That is, using augmented vectors, we would ideally like $\mathbf{w}'\mathbf{x}_i > 0$ for class 0 design set elements and $\mathbf{w}'\mathbf{x}_i \leqslant 0$ for class 1. Here the i subscript refers to the ith design set point. In fact a historically important approach simplified things by the following artifice (which does not generalise to more than two classes). Define $\mathbf{y}_i$ as $\mathbf{y}_i = \mathbf{x}_i$ for points from class 0 and $\mathbf{y}_i = -\mathbf{x}_i$ for points from class 1. Now, in terms of the $\mathbf{y}_i$, we would ideally like the classification rule to yield $\mathbf{w}'\mathbf{y}_i \geqslant 0$ for *all* points in the design set. If we could achieve this, then no design set points would be misclassified.

The *perceptron criterion* for choosing the weights is based on this principle. It is defined as $-\sum_M \mathbf{w}'\mathbf{y}_i$, where the summation is over the points which are misclassified by the weight vector $\mathbf{w}$. Points which are incorrectly classified contribute an amount which depends on their distance from the decision surface $\mathbf{w}'\mathbf{y} = 0$. Points which are correctly classified do not figure in this measure, no matter how close they are to the decision surface. The perceptron criterion is an attractive one because it is continuous. There are no sudden jumps as the weight vectors change and points cross from one side of the decision surface to the other—as there are, for example, with straightforward error rate. It is also piecewise linear in the weight components. These properties make it amenable to various optimisation methods and many such methods can be and have been applied. An obvious choice, since the perceptron criterion is continuous, is steepest descent.

The gradient vector of the criterion is $-\sum_M \mathbf{y}_i$ so that the $(l + 1)$th updating step has the form $\mathbf{w}_{l+1} = \mathbf{w}_l + \rho_l \sum_{M_l} \mathbf{y}_i$, where ρ_l is a function of l which determines step size. M_l is the set of points which are misclassified at the lth step. It is not difficult to see that if the classes are linearly separable then this procedure is guaranteed to converge to a solution (and again we caution that we are not at present considering generalisability issues). This steepest descent approach is very much a mathematician's or statistician's strategy for choosing an optimising set of weights (even if the criterion itself is very much a computer scientist's). A variant of the steepest descent method more in tune with computational styles of thinking is to present the points one at a time to the current classification rule (i.e. to classify each point using the current weight vector), leaving this vector as it is when a point is correctly classified, and updating the weight vector when a point is incorrectly classified, using the updating $\mathbf{w}_{l+1} = \mathbf{w}_l + \rho_l \mathbf{y}_i$. For obvious reasons, such approaches are called *error correction procedures*. The design set is repeatedly cycled through, adjusting the weights whenever a point is misclassified, and again it can be shown that, for suitable ρ_l, a separating surface will be found if the points are linearly separable (for example, if $\rho_l = 1$ for all l).

Since $-\sum_M \mathbf{w}'\mathbf{y}_i$ is piecewise linear in the weights, another optimisation strategy which suggests itself is linear programming. Some adjustments must first be made to recast the problem into standard linear programming format. First, the criterion $-\sum_M \mathbf{w}'\mathbf{y}_i$ is non-negative for all choices of $\mathbf{w}$, so it would be minimised by $\mathbf{w} = \mathbf{0}$. This is avoided by imposing a constraint on the components of $\mathbf{w}$. The obvious constraint $\mathbf{w}'\mathbf{w} = 1$ is nonlinear so, instead, the requirement $\mathbf{w}'\mathbf{y}_i \geq 0$ for points in the design set is replaced by the requirement $\mathbf{w}'\mathbf{y}_i \geq b_i$ for a set of values $b_i > 0$. Any solution to the adjusted problem is certainly also a solution to the original problem. This adjustment introduces a safety margin: we are now trying to find a decision surface which not only classifies all points in the design set correctly but for which moreover the ith point is no closer than $b_i/\sqrt{\mathbf{w}'\mathbf{w}}$ to it. Secondly, the criterion is *piecewise* linear, rather than merely linear. A standard technique exists for converting from the piecewise linear to the latter. Hand (1981a, Section 4.3) describes in detail the whole process of optimising the perceptron criterion by linear programming and presents an example.

The perceptron criterion $-\sum_M \mathbf{w}'\mathbf{y}_i$ can also be written as $U_1 = -\sum_i (1 - g(\mathbf{w}'\mathbf{y}_i))\mathbf{w}'\mathbf{y}_i$, where g is the indicator function—or *threshold logic unit* (TLU) in computer science terminology—taking the value -1 if the argument is negative and $+1$ otherwise. Alternatively, without the sign change introduced between $\mathbf{x}$ and $\mathbf{y}$ (which was useful for the special case of two classes) we can put $U_1 = -\sum_i (b_i - g(\mathbf{w}'\mathbf{x}_i))\mathbf{w}'\mathbf{x}_i$, defining b_i as $+1$ for class 0 points and -1 for class 1 points. The summation here is over *all* design set points. The first factor in the argument becomes zero when points are correctly classified: the argument of g having the same sign as b_i in such cases. Two observations can be made about U_1 when it is expressed in this form. First, we see that it is a distance measure between the vector $\mathbf{b}$ and the vector of predicted values $(g_1(\mathbf{w}'\mathbf{x}_1), \ldots, g_1(\mathbf{w}'\mathbf{x}_n))'$. This will allow us to make obvious generalisations to multiclass situations and to use alternative distance measures. Secondly, $g(\mathbf{w}'\mathbf{x})$ has the form of a generalised linear model (GLM), with g the inverse link function. Obviously, what we are talking about here are standard statistical models approached from a different direction.

The 'generalised' in the GLM of the perceptron comes from the nonlinearity of the g indicator function. What if we do not generalise? What if we merely take a linear model? That is, we wish to find a weight vector $\mathbf{w}$ which minimises some measure of the difference between $\mathbf{b}$ and $(\mathbf{w}'\mathbf{x}_1, \ldots, \mathbf{w}'\mathbf{x}_n)'$. If we adopt, as our distance measure, the sum of squared differences between the components of these two vectors, we have a standard least squares linear regression problem. In this case the vector of differences $(\mathbf{b} - (\mathbf{w}'\mathbf{x}_1, \ldots, \mathbf{w}'\mathbf{x}_n)')$ can alternatively be written as $(\mathbf{b} - \mathbf{X}\mathbf{w})$ where the ith row of $\mathbf{X}$ is $\mathbf{x}'_i$. The criterion to be minimised is then $U_2 = (\mathbf{b} - \mathbf{X}\mathbf{w})'(\mathbf{b} - \mathbf{X}\mathbf{w})$ with solution $\hat{\mathbf{w}} = (\mathbf{X}'\mathbf{X})^{-1}\mathbf{X}'\mathbf{b}$. This is a standard statistical solution. An alternative solution developed in the perceptron criterion community (and motivated by their interest in adaptive, sequential 'learning' methods) assumed that the design set points are presented to the system one at a time, so that it must update the weight vector at each step. If we were to use steepest descent with U_2 above, we would take steps in the direction $-\partial U_2/\partial w_i = 2\sum_i (b_i - \mathbf{w}'\mathbf{x}_i)x_{il}$. However, in the spirit of the sequential approach, let us cycle through the design set points one at a time. The updating rule then becomes $\mathbf{w}_{(k+1)} = \mathbf{w}_{(k)} + 2\rho(b_i - \mathbf{w}'\mathbf{x}_i)\mathbf{x}_i$. This is called the *Widrow–Hoff rule*, the *delta rule* or the *adaline rule* (from adaptive linear). It is an adaptive way of finding the least squares solutions. It is perhaps worth remarking here that the different algorithms—the linear algebra regression solution and the adaptive method—have properties which are to some extent complementary. For example, the former may be quicker but the latter corrects any rounding errors as new points are introduced. This method has an essential difference from the perceptron error correction procedure because all points lead to an adjustment (assuming $(b_i - \mathbf{w}'\mathbf{x}_i)$ is never exactly zero); in the perceptron procedure it is only misclassified points that lead to an adjustment. It also follows that, even if the samples are perfectly separable, this least squares criterion approach may converge to a local minimum not in the class of solution vectors $\mathbf{w}$.

So far we have considered two forms for the *activation function g*, forms which we might regard as being at opposite extremes. These are the indicator function version and the identity function version. An attractive intermediary is to consider some kind of sigmoidal function, with a parameter determining its steepness. For large parameter values it will then approach one of the above extremes and for small parameter values it will approach the other. Moreover, if we adopt a smooth function, its differentiability properties may be useful. A very common such function in these applications is the logistic function $g(\mathbf{w}'\mathbf{x}) = [1 + \exp(-2\beta\mathbf{w}'\mathbf{x})]^{-1}$ and another which is sometimes used is $g(\mathbf{w}'\mathbf{x}) = \tanh(\mathbf{w}'\mathbf{x})$. In general such gs lead to learning steps $\rho(b_i - g(\mathbf{w}'\mathbf{x}_i))\partial g/\partial \mathbf{w}$ (cf. the score function in estimating the parameters of GLMs).

The sum of squared distance (also called the quadratic distance, the Euclidean distance or sometimes the Pythagorean distance) between $\mathbf{b}$ and $(\mathbf{w}'\mathbf{x}_1, \ldots, \mathbf{w}'\mathbf{x}_n)'$ is not the only distance measure which may be used, though it has been explored most intensively. Another obvious choice (at least, from a statistician's perspective) is the log-likelihood:

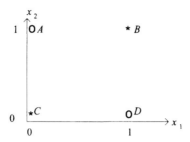

Figure 3.4   No matter where a linear decision surface is positioned, it cannot separate the class of stars (points B and C) from the class of circles (points A and D).

$$\sum_i [c_i \ln g(\mathbf{w}'\mathbf{x}_i) + (1 - c_i)\ln(1 - g(\mathbf{w}'\mathbf{x}_i))]$$

where $c_i = (1 - b_i)/2$, i.e. c_i is 0 for objects from class 0 and 1 for objects from class 1. More generally still, the performance measures described in Chapter 6 can be used.

The most comprehensive study of simple single-layer perceptrons was that of Minsky and Papert (1969) who showed that such architectures had severe limitations. The simplest example of such a limitation is shown in Figure 3.4. It is obvious that a *linear* surface, of the kind computed by a single-layer machine, cannot separate the two classes perfectly. The way to overcome this problem had been known for some years before this—introduce more layers of processors. Such structures are described by Palmieri and Sanna (1960) and Gamba *et al.* (1961).

3.3.3   *Multilayer perceptrons*

Early work on statistical pattern recognition often drew a distinction between *feature extraction*—the process of combining and transforming the raw variables to provide input to a classification rule—and *constructing the classification rule* itself. The distinction may sometimes be helpful for practical reasons, but it is rather artificial. Both steps are really aspects of the construction of a flexible decision surface. However, it can be useful when we begin to consider more general types of feedforward neural network; these can be thought of as a sequence of steps involving combining and transforming outputs from earlier stages. So the simple perceptron above consisted of a linear combination of the raw variables. (The subsequent nonlinear transformation by g is, in a sense, irrelevant in the simple perceptron; its monotonicity means that linear decision surfaces remain linear.) Now let us extend it to be a *linear combination of* nonlinear transformations of linear combinations of the raw variables. That is, we are inserting several linear combination stages (just two in this example), with the output of each stage undergoing a nonlinear transformation. Put another way, we are inserting extra *layers* of nodes in the network. The simple perceptron is often called a *single-layer network*, with the 'input layer' being numbered 0. Clearly, in generalising to more layers, the nonlinear transformations will play a fundamental role. Without such transformations we have linear combinations of linear combinations—and these are equivalent simply to linear combinations. This means that the region of the measurement space in which such a sum has value greater than some threshold must be determined by a simple hyperplane (a line in two dimensions). In contrast, a sum of nonlinear functions is generally nonlinear. For example, consider the sum of x^2 and $(1 - x)^2$. This is also a quadratic function, taking its minimum at 1/2. Defining a classification region by comparing this with a threshold would induce a central region to be classified as one class, surrounded by two regions belonging to the other class. That is, by virtue of the nonlinear transformation, we can generalise from simple linear decision surfaces.

Mathematically, with just two stages, we can express this as $\sum_k v_k y_k = \sum_k v_k g_k \{\sum_l w_{kl} x_l\}$. Here the g_k are nonlinear transformations and the $y_k = g_k \{\sum_l w_{kl} x_l\}$ are new 'derived' variables, or features. To illustrate how this overcomes the problem noted at the end of the preceding section, let $\mathbf{x}' = (1, x_1, x_2)$ and suppose that the g_k are indicator functions (or TLUs). That is $g_k(\mathbf{w}_k' \mathbf{x}) = 0$ if $\mathbf{w}_k' \mathbf{x} < 0$ and $g_k(\mathbf{w}_k' \mathbf{x}) = 1$ otherwise. Define two 'derived' variables as follows:

$w_{10} = -1/2$, $w_{11} = 1$, and $w_{12} = 1$ so that

$$y_1 = g_1 \left\{ -\frac{1}{2} + x_1 + x_2 \right\} \text{ which is 0 if } x_1 + x_2 < 1/2 \text{ and 1 if } x_1 + x_2 \geqslant 1/2$$

$w_{20} = -3/2$, $w_{21} = 1$, and $w_{22} = 1$ so that

$$y_2 = g_2 \left\{ -\frac{3}{2} + x_1 + x_2 \right\} \text{ which is 0 if } x_1 + x_2 < 3/2 \text{ and 1 if } x_1 + x_2 \geqslant 3/2$$

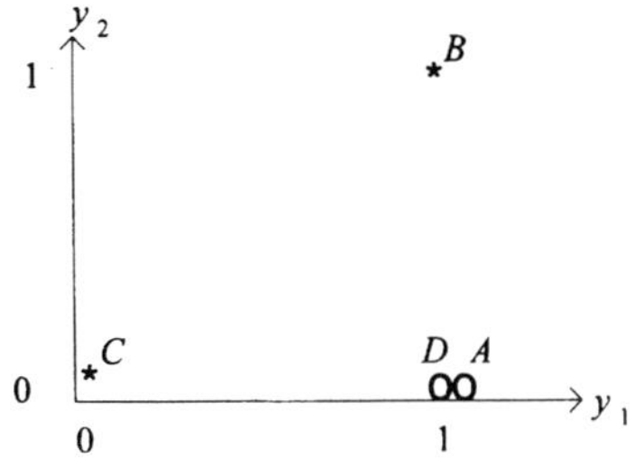

Figure 3.5 In the space spanned by the new variables y_1 and y_2 it is easy to separate the class of stars from the class of circles.

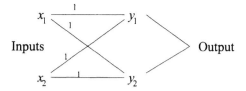

Figure 3.6 The network representation of the transformations from the x variables to the y variables and the output.

Figure 3.5 shows the positions of the four objects in Figure 3.4 when they are plotted using the new variables y_1 and y_2 as axes. We see that in this (y_1, y_2) space the points may be separated by a simple linear surface. That is, having taken linear combinations of the raw variables and having transformed them by nonlinear transformations, we can now take a linear combination $(\sum_k v_k y_k)$ of the results to yield a decision surface which will separate the objects.

Figure 3.6 shows what this looks like in the neural network graphical representation. Instead of the input nodes feeding straight through to a single output node, there is now an intermediate layer of nodes, corresponding to the two new derived variables. The weights on all of the links are unity, but the thresholds for the y variables differ.

The simple perceptron can be generalised to more than two classes by having an output node for each class, with all of the input nodes being connected to each of them. The links, each with their own set of weights, will form a linear combination of the inputs for each output. A similar generalisation applies to multilayer perceptrons.

In the above example, we chose the weight vectors so they would partition the design set of Figure 3.4 in the way we wanted. This was feasible since the measurement vectors had only two components. More generally, however, multiple measurements will be involved and we will want some kind of automatic way to estimate the weights. The difficulty of finding an algorithm to do this was one of the factors which slowed down research in this area, even though the ideas of multilayer systems had been introduced as early as 1960. Minsky and Papert (1969), writing of multilayer systems, observe (p.206): 'It ought to be possible to devise a training algorithm to optimise the weights in this using, say, the magnitude of a reinforcement signal to communicate to the net the cost of an error. We have not investigated this.' In fact, although effective algorithms for estimating the weights had been invented several time previously, it was not until the 1980s that the possibilities they offered were widely appreciated (coupled with the practical implications of great advances in computing hardware).

The algorithm which made progress practicable was the *back propagation* algorithm—a steepest descent method applied to the layers of weights in a multi-layer perceptron. To give an outline, let us generalise the above two-layer single output node network to multiple output nodes. (The generalisation to more than two layers is straightforward and algebraically rather tedious.) Input nodes are indexed by l, hidden nodes are indexed by k, and output nodes are indexed by j. We then have

$$z_j = g_j\left\{\sum_k v_{jk} y_k\right\} = g_j\left\{\sum_k v_{jk} g_k\left\{\sum_l w_{kl} x_l\right\}\right\}$$

with input nodes x_l, weights w_{kl} connecting the lth input node to the kth hidden node (y_k), and weights v_{jk} connecting the kth hidden node to the jth output node (z_j).

We will have a training set from which to derive our estimates of the weights (parameters) of the system. As usual, this will consist of a sample of individuals for which we know both the measurement vectors ($\mathbf{x}_i$) and the true classes. We let $\mathbf{c}_i$ denote a vector consisting of 0s apart from a single 1 in the jth position, where j is the true class of the ith individual. Adopting the least squares criterion, we seek that set of weights which minimises

$$E = \sum_i \sum_j (c_{ij} - z_{ij})^2 = \sum_i \sum_j \left[c_{ij} - g_j\left\{\sum_k v_{jk} g_k\left\{\sum_l w_{kl} x_{il}\right\}\right\}\right]^2$$

where the summation over i is over the elements of the training set. Provided the nonlinear transformations g are differentiable, we can use standard steepest descent methods. (Just why it should then be called back propagation we will see in a moment.)

First let us see what steps the steepest descent method requires us to take in the directions spanned by the weights connecting the hidden nodes to output nodes. We have

$$\frac{\partial E}{\partial v_{jk}} \propto \sum_i (c_{ij} - z_{ij})\frac{\partial z_{ij}}{\partial v_{jk}} = \sum_i (c_{ij} - z_{ij})g_j' y_{ik} = \sum_i \delta_{ij} y_{ik} \qquad (3.1)$$

where

$$\delta_{ij} = (c_{ij} - z_{ij})g_j' \qquad (3.2)$$

with g_j' indicating the derivative of g_j with respect to its argument and y_{ik} signifying the value of the kth hidden node for the ith training set element. For the input node to hidden node links we have

$$\frac{\partial E}{\partial w_{kl}} \propto \sum_i \sum_j (c_{ij} - z_{ij})\frac{\partial z_{ij}}{\partial w_{kl}} = \sum_i \sum_j (c_{ij} - z_{ij})g_j' v_{jk} g_k' x_{il}$$
$$= \sum_i \sum_j \delta_{ij} v_{jk} g_k' x_{il} = \sum_i \delta_{ik} x_{il} \qquad (3.3)$$

where

$$\delta_{ik} = \sum_j \delta_{ij} v_{jk} g'_k = g'_k \sum_j \delta_{ij} v_{jk} \qquad (3.4)$$

We see that the updating steps connecting both pairs of adjacent nodes have the same forms. In general, for a network with an arbitrary number of hidden layers, the updating steps for the weights connecting the mth and $(m + 1)$th layers (taking the input layer as the 'first' layer and the output as the 'last') will be proportional to $\sum_i \delta_{i(m+1)} u_{im}$, where the u are the values on the input nodes to the links in question.

Again, as with the simple perceptron outlined in the preceding section, in many applications the updating is performed sequentially, with training points being presented one at a time (on line). This sequential approach is probably more in tune with the computational orientation to estimation, with 'batch' presentation of all the points at once, more in tune with the statistical orientation. There is some evidence that the batch strategy has more difficulties with local minima.

Equations (3.1) – (3.4) show us why this algorithm is called backpropagation. The δ in the input node to hidden node links (3.4) are formed by combining the δ in the hidden node to output node links (3.2). That is, the errors are calculated by propagating them backwards. So, to implement the method in practice, one begins with a random choice of weights and propagates the input xs through the network using these weights. When output 'predicted values' have been reached, one calculates the δ for the output layer and back-propagates them through the network. The resulting δ are then used to update all the links. This whole procedure is then repeated for a new element of the training set. One can see from this why the method is not very fast. If the logistic function $g(s) = [1 + \exp(-2\beta s)]^{-1}$ is used as the activation function, then (3.2) can be written as $\delta_{ij} = (c_{ij} - z_{ij}) 2\beta g(s)(1 - g(s))$, a form which is sometimes presented.

3.3.4 How many layers?

A network with no hidden layer induces a single hyperplane in the measurement space. Only sets of points which are linearly separable can be perfectly correctly classified by such a system. Introducing a hidden layer, and permitting nonlinear transformations (so that the whole does not reduce to the equivalent of no hidden layer), results in a system which will have a function value above a threshold in a convex region of the measurement space. (It carries out an AND operation; for example, if functions a and b in the previous layer are above a threshold, this node will give a positive output.) Finally, introducing a further layer allows such convex regions to be combined, producing nonconvex, even disconnected, regions if necessary. Thus, in principle, two hidden layers are sufficient for any problem. In practice, it may be advantageous to use more than two layers. Moreover, increasing the complexity of the nodes can have dramatic advantages.

     * * oo   *   *

Figure 3.7   In this one-dimensional example, with variable x, the class of stars cannot be separated from the class of circles using a simple linear surface (which is equivalent to a point in one dimension). However, the classes can be separated if x^2 is also used (here this corresponds to two points).

Earlier we described the view of a sequence of layers inducing a sequence of transformations which successively distort the measurement space until one ends up with a representation that permits linear separability. Indeed this is illustrated in the example of Figure 3.4, solved by introducing an extra layer. Each node in such an intermediate layer may be regarded as extracting a *feature* from its input layer. In general one may combine and transform outputs from the preceding level. This, in a slightly different guise, is a concept which will be familiar to all statisticians. A very simple example is illustrated in Figure 3.7. The two classes are not linearly separable in terms of x alone, but can be linearly separated in the space spanned by x and x^2.

This notion of transforming outputs from the preceding layer permits models of arbitrary complexity to be constructed. We could, for example, define transformations yielding terms from an orthonormal basis function expansion. Thus neural networks can be viewed as a completely general function estimation methodology.

3.4   GENERALISED ADDITIVE MODELS

Kernel and nearest neighbour methods model the functions $f(j|\mathbf{x})$ over the multivariate $\mathbf{x}$ space, essentially using local models. In contrast, linear methods form a single global model. Compromises between these two extremes are possible. *Additive models* are one such. In place of the linear model form $f(j|\mathbf{x}) = \alpha_j + \sum_k \beta_{jk} x_k + \varepsilon$ they take the form $f(j|\mathbf{x}) = \alpha_j + \sum_k g_{jk}(x_k) + \varepsilon$. That is, more general transformations of the raw variables are permitted, beyond a simple multiplication by a constant.

Suppose that all but one of the variables in $\mathbf{x} = (x_1, \ldots, x_d)$ are fixed. Then, in a linear model, the effect on $\hat{f}(j|\mathbf{x})$ of changing the outstanding x_k will be independent of the values at which the others are fixed. Precisely the same is true of additive models. This decoupling substantially aids interpretation. However, whereas in a linear model the form of the relationship between each x_k and $\hat{f}(j|\mathbf{x})$, with the other variables being fixed, is highly constrained (to a simple linear function), for additive models the form is much more general. This means that additive models are more flexible and so might be expected to fit the data better. On the other hand, because they transform each variable separately, they are not as flexible as a method which takes the multivariate nature into account.

We need to consider what sort of transformations to use and, since they will typically have free parameters (like the coefficients in the linear model), how to estimate them. If we use standard families of transformations (such as log, square root, etc.), the model will be standard multiple regression, simply applying it to transformed variables. Or we could include the exponents of the x_k as parameters to be chosen (so that, once we have worked out how to estimate the parameters, the system automatically chooses from the set of power transformations).

Taking this further, we could let g be a polynomial in x_k, with several coefficients (and exponents) to be chosen. And even more generally, g could be described in terms of any set of basis functions. This is equivalent to replacing the summation over the variables in $f(j|\mathbf{x}) = \alpha_j + \sum_k g_{jk}(x_k) + \varepsilon$ by a summation over a larger set of transformed versions of the variables. Each of the raw variables is used to generate (by transformation) a set of new derived variables (or features, as they are called in the pattern recognition literature). Of course, we will want to avoid having too many such functions, or else we will need a vast design set. Splines are a popular choice of basis function. These are functions comprised of segments which are polynomial of some degree b between specified values of x_k (called knots) and which are differentiable of order $b - 1$ at the knots. The overall function is thus piecewise polynomial. The flexibility of the function is determined by the order of the polynomial segments and the number of knots. Cubic polynomials are a common choice. Parameter estimation in methods using basis functions can be via ordinary regression on the extended variable space, although in the spline case the knots must be chosen first. They can be distributed uniformly over the range or placed at the quantiles of the range of the design set values of x_k.

There are many variants of the spline approach. An important one is multivariate adaptive regression splines (MARS), which uses multivariate splines as the basis (formed as products of the univariate spline basis functions).

An alternative to spline type models is to use a univariate kernel smoother over x_k to provide weights which are used to average the classes of the design set elements. More generally, the kernel weights can be used to fit a local low-order (e.g. linear) polynomial. The *loess* method uses this sort of principle, but based on the k nearest neighbours rather than a kernel.

Of course, there is no reason (apart from the increase in computational cost and in complexity, which means more design data are required) why these ideas should not be extended to include transformations of more than one variable at a time. In the limit, with all of the variables included in a single g function, we are fitting a model to the whole space. For example, an overall kernel method is of this form. Of course, including more than one variable in a single g function means that the convenience of the decoupling is sacrificed.

All of these methods have parameters which need estimating. A standard approach is the *backfitting* method. This is an iterative approach. For two classes, starting with initial values for the parameters, the method cycles through each of $x_1, \ldots, x_d$ in turn, fitting models excluding each of them. For a model not including x_l, the residuals are $c_i - \alpha - \sum_{k \neq l} g_k$ where c_i is the true class of the ith design set object. The method finds those parameters for the g_l function which most accurately predicts these residuals. This is then repeated for each of the other x_k until the system converges.

Linear models are all very well when attention is focused on the decision surface, but when attention is focused on the conditional probabilities $f(j|\mathbf{x})$ they leave something to be desired. We have already remarked that they can, for example, lead to probability estimates outside the interval $[0, 1]$. Models such as logistic regression overcome this by matching the linear form to a transformed response. Precisely the same sort of generalisation can be made for additive models, yielding models of the form $h(\mu) = \alpha + \sum_k f_k(x_k)$, where μ is the expected true class value.

3.5 PROJECTION PURSUIT REGRESSION

We commented in the previous section that, in principle at least, additive models could be extended to include more than one variable (even all the variables) in each component g function. One variant of this idea is *projection pursuit regression*, which lets g transform a linear combination of the variables. We have

$$f(j|\mathbf{x}) = \alpha_j + \sum_r g_{jr}(\beta'\mathbf{x}) + \varepsilon$$

This is a sum of transformations of linear combinations of the raw x variables. The idea behind such models is easiest to grasp in the case of two classes with a single transformed linear combination. In this case, the linear combination defines a direction in the x space. The 'transformation' may be a simple smoothing in this direction, as with additive models. Thus this approach may be regarded as a generalisation of additive models in which the additive components are not restricted to the directions of the variables (but not so great a generalisation that arbitrary combinations of variables are permitted).

In the case of a single component, the estimated probabilities will provide constant contours in directions orthogonal to the specified direction β. However, the gaps between these contours will depend on the g function. (There is a similarity to the model described in Section 5.3 where the direction β is chosen by fitting a classical linear discriminant function, but the actual classifications are then made using nearest neighbour methods in a slight modification of the space defined by β.) With more than one g component, the model is not so readily interpreted, although one can think of each linear combination as a new derived feature and the overall model as a classification rule built using these features. The similarity will be obvious between this approach and the nested series of linear combinations of nonlinear transformations of the raw variables used in feedforward neural networks.

3.6 RADIAL BASIS FUNCTIONS

The methods described above, having (at least, as part of their model) the form of a sum of transformations of the raw variables, can be regarded as sums of basis functions. A different kind of basis is given by *radial basis functions*.

Radial basis functions, mixture decompositions and kernel methods can all be viewed as variants of the same idea. The description *potential function method* is often used as a general term for this broad class of methods. A mixture model for a density function $f(\mathbf{x})$ has the form $f(\mathbf{x}) = \sum_{r=1}^{M} p_r f_r(\mathbf{x})$, where the $f_r(\mathbf{x})$ (the basis functions) are distributions from a specified family (requiring the location and perhaps other parameters to be given) and the p_k are weights (or prior probabilities) giving the relative contributions of the kth class to the mixture. A very simple special case is when the $f_r(\mathbf{x})$ are multivariate normal distributions, the rth one centred at some point $\mathbf{x}_r$, and all with identity covariance matrix. The trick in fitting such models is to find good values for the $\mathbf{x}_r$. If, as in this example, each basis function is solely a function of distance from its centre $\mathbf{x}_r$, it is called a radial basis function. The kernel method, described in Chapter 5, takes this to the limit and uses one component centred at each design sample point.

This sort of approach can also be used to estimate the $f(j|\mathbf{x})$ directly, regarding the separate components as basis functions. Common choices include $g(s) = e(-s^2/2\sigma^2)/\sigma$, $g(s) = s^2 \log s$, $g(s) = (1 + s^2)^{-1}$ and $g(s) = (c^2 + s^2)^{1/2}$, and extensions permit aspects of the basis function to vary from location to location, perhaps by permitting the σ to vary from function to function in $g(s) = e(-s^2/2\sigma^2)/\sigma$.

Vector quantisation is also a closely related approach. Here the measurement space is 'quantised' or partitioned according to proximity to one of a number of 'principal points'. Some of these points will belong to each class, so permitting classification on a single nearest neighbour approach. This classification strategy is common in speech recognition (Section 10.4).

3.7 FURTHER READING

One of the first descriptions of logistic regression is to be found in Berkson (1944) and an excellent introductory account is given in Collett (1991). For the multivariate normal case with equal covariance matrices, when both logistic regression and Fisher's linear discriminant function yield the optimal decision surface, Efron (1975) shows that logistic regression, using only the information in the conditional probabilties for given $\mathbf{x}$ is less efficient than linear discriminant analysis.

Hand (1981a, Chapter 4) describes the basic perceptron in some detail, illustrating various optimisation methods. The perceptron model was originated by F. Rosenblatt (1962) and convergence proofs may be found in Nilsson (1965). Similar structures and methods, including the Widrow–Hoff rule, were also described by Widrow and Hoff (1960) and Widrow (1962). Methods for minimising the sum of squared differences by choosing both **w** and **b** are described by Duda and Hart (1973) and Ho and Kashyap (1965, 1966). The key book on perceptrons is Minsky and Papert (1969).

Amongst the first descriptions of multilayer perceptrons were those of Palmieri and Sanna (1960) and Gamba *et al.* (1961). Their form used an indicator function as the nonlinear transformation. The invention of the back propagation learning rule was one of the main advances that rekindled interest in neural networks. It was invented independently by several different researchers, including Bryson and Ho (1969) and Rumelhart *et al.* (1986). Adaptations and alternatives to the back propagation algorithm have been explored by many authors, including Plaut *et al.* (1986), Solla *et al.* (1988), Jacobs (1988), Fahlman (1989) and Rohwer (1990). More generally, a vast amount of work on general optimisation theory and methods has been carried out by mathematicians and allied researchers, and this can usefully be brought to bear in developing improved algorithms.

There are now a large number of books discussing neural networks. Particularly noteworthy are Hertz *et al.* (1991) and Bishop (1995). Additive models and generalised additive models are described in Hastie and Tibshirani (1986, 1990). Projection pursuit regression is described in Friedman and Stuetzle (1981) and multivariate adaptive regression splines are described in Friedman (1991). Potential function methods are described in Coomans and Broeckaert (1986) and a good review of vector quantisation methods is given in Gersho and Gray (1992).

CHAPTER 4

Recursive Partitioning Methods

4.1 INTRODUCTION

Fisher's linear discriminant method, discussed in detail in Chapter 2, is probably the oldest formal statistical classification method in widespread use. However, it is arguable that recursive partitioning methods, or *tree* classifiers, are the oldest conceptually, since the basic idea is so straightforward. The underlying idea of tree classifiers has cropped up in numerous different forms, typically with the researchers in one field being unaware of parallel or earlier work in other areas.

Prior to the widespread availability of computers the construction of classification trees used *ad hoc* methods, based on the knowledge or opinion of experts. In some contexts such informal methods are still used, but nowadays it is also possible to apply formal statistical model building and estimation methods in an attempt to find the 'best' tree structure using a design set of objects ('best' being defined in some suitable way). The bulk of this chapter describes such efforts.

Three areas of especial note in the early development of tree construction are numerical taxonomy, medicine and the social sciences. We shall say a little about each of these areas below. More recently, researchers in both statistics and artificial intelligence (machine learning, pattern recognition, etc.) have made substantial advances, initially in parallel without apparently being aware of what was going on in the other discipline, but during the last decade or so, in a collaborative mode. As is discussed further in Chapter 11, the statistical and artificial intelligence schools tackle the same problem, but emphasise different aspects. For example, statisticians, seeing things in terms of probability distributions, expect to find overlapping distributions and often think in terms of continuous variables. In contrast, the machine learning community stress perfect separability and often think in terms of discrete, often binary, variables and their Boolean combinations.

We begin by outlining the simplest form of tree classifier.

Figure 4.1 illustrates an imaginary tree constructed for classifying hospital patients into one of two disease classes. To classify a new patient, at the top or

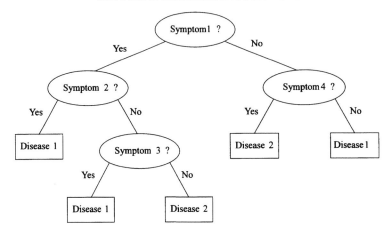

Figure 4.1 An imaginary tree for classifying a patient into one of two disease classes according to their symptom pattern.

root node of the tree, one first notes whether the patient shows symptom 1. If the symptom is present one looks for symptom 2. If symptom 1 is absent one looks for symptom 4. In this manner one works down the tree, *recursively* applying the same kind of operation: look for a symptom and decide which branch to follow on the basis of whether that symptom is present or absent. Ultimately one reaches a *leaf* node, beyond which one cannot go. Each of these leaf nodes is associated with a particular disease class; a patient traversing the tree and arriving at a node associated with disease class 1 will be diagnosed as suffering from disease 1.

In this tree structure each nonterminal node is associated with a single variable and a partition of that variable into two classes determines which branch a new patient follows. In the example in Figure 4.1, each of the variables is binary, just permitting answers 'yes' or 'no', but in general the variables may be categorical with more levels or may even be continuous. Then part of the problem of designing the tree is deciding how to partition each variable. There is no reason why a variable should not reappear further down the tree (associated with a different partition, otherwise it would not produce a split). This fact allows us to show, in two dimensions, an alternative way of visualising a tree classifier: in terms of the space spanned by the measured variables. So, Figure 4.2(a) shows another simple tree classifier, this time involving only two variables. First, variable x of the new object to be classified is examined to see if it is greater than 1. If it is less than 1 then variable y is examined. According to whether y is less than or greater than 1, the object is assigned to class 1 or class 2. Following the other branch of the tree, if x is greater than 1, a test is made to see if it is greater than 1.5. If x is greater than 1.5, the object is assigned to class 1, otherwise the object is assigned to class 2. Figure 4.2(b) shows the measurement

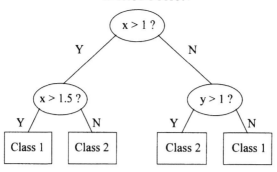

Figure 4.2(a) A simple tree classifier with two continuous variables.

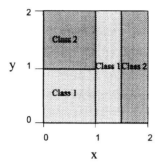

Figure 4.2(b) The measurement space representation of the classifier in Figure 4.2(a).

space representation. The space is partitioned into cells, each of which corresponds to a particular class of object.

These simple examples illustrate the use of a tree, but the question then arises as to how the tree should be built in the first place. Which variable should we use at each node? What partition of that variable should we use? Should the partition of each node be based on more than one variable—a combination of variables instead of just one variable—i.e. should the partitions be *polythetic* instead of *monothetic*? If so, what combination? Which class should each leaf node correspond to? And so on.

There are also other, less obvious, issues raised by these examples. Figure 4.2(b) shows an essentially continuous underlying measurement space being partitioned and we can see that the boundaries of the cells are orthogonal to the variables. Is this restrictive (for example, what if the true decision surface is diagonal)? If it is, how can it be overcome? For nominal (discrete, unordered) variables such partitions might seem to be very natural. And the emphasis of each research community again becomes relevant; machine learning theorists focus on discrete variables, statistical theorists on continuous variables.

An associated difference is that machine learning theorists sometimes see the construction of a tree as but a stage towards the final objective, which is to describe each nodal partitioning as an antecedent–consequent rule in an expert system. That is, the hierarchical tree structure is re-expressed as a uniform collection of simple if–then rules. Such collections of rules are thought, at least according to some theories, to provide good models of the way human brains hold procedural knowledge. The emphasis on categorical data leads to such rules having simple Boolean antecedents. An example is given by Michalski and Chilausky's (1980) work on soybean pathology. The rules they extracted from the tree include

IF (*leaf malformation = absent* AND *stem = abnormal* AND *internal discolouration = black*) THEN (*diagnosis = charcoal rot*)

IF (*leaves = abnormal*) AND (*leaf malformation = absent*) AND (*leaf mildew growth = on upper leaf surface*) AND (*roots = normal*) THEN (*diagnosis = powdery mildew*)

Michalski and Chilausky compared the performance of an expert system using if–then rules derived from the recursive partition of a training set with the performance of a system using rules derived by consultation with human experts in the area. Contrary to their initial expectation, the former performed better. They suggest several possible reasons for this, including the fact that the experts, though expert in *making* the diagnoses, may not have been expert in *explaining how* they made those diagnoses.

Classification trees are often described as nonparametric since they do not assume any underlying family of probability distributions. They are also very flexible; for example, they handle interactions between the variables with ease. Another merit is their conceptual simplicity. This is no mean advantage since in practical applications it is often vital that a client, perhaps a doctor, can obtain an intuitive grasp of how a decision is reached and have confidence in the methodology. (In other applications, such as automated speech recognition, this property is less important. This illustrates the fact that choice of classification method depends on the application.) Trees can also handle mixed variable types with consummate ease, and can cope with measurement vectors of different dimensionality from object to object, a property not possessed by all other classification methods.

Both of the examples above show more than one leaf node for each class. This is perhaps typical of most applications of tree classifiers, which partition the measurement space into disconnected regions. However, some problems involve a very large number of classes (recognising Chinese characters is a problem with thousands of classes). In such cases it might be more natural to try to group classes into 'superclasses' at high levels of the tree, splitting them up lower down, and arranging things so that ultimately there is only a single leaf node for each class. Again, different problems will require different kinds of solutions; even within a given class of methods there are different variants.

4.2 THREE IMPORTANT APPLICATION AREAS

4.2.1 Numerical taxonomy

Different disciplines have their own terminology. So that which, in statistics, is called a *variable* is called a *character* in taxonomy. The levels of a (nominal) variable are called *states* or *attributes*. The classes into which objects are to be allocated are called *taxa*. In numerical taxonomy an important property of a potential character is *ease of observation*. This means that they may not be fundamental or even 'best' in the sense of providing maximal discriminatory power, but that the values of the attributes can be identified quickly and easily.

Examples of taxonomic trees are given in Payne and Preece (1980, Figure 1) for identifying common British trees, and Pankhurst (1991, p.6) for identifying British *Epilobium* species.

4.2.2 Medicine

There is evidence suggesting that human experts approach medical diagnosis in a hypothetico-deductive way (e.g. Hand, 1985, Section 5.2). That is, presented with a patient, they come up with some possible explanations for the symptoms and then see if further facts conform to those tentative diagnoses. This is clearly very different from the sequential style of classification trees, incrementally collecting information until a decision is made, with no hypothesis generation stage. However, this does not mean that diagnostic systems built using tree approaches may not outperform alternatives. Moreover, as a prescription of a diagnostic process to be followed by someone else (possibly not an expert in the area) classification trees are hard to beat. In any case, although the hypothetico-deductive approach may be ideal in some circumstances, it will not do in others. For example, it relies on the user generating the 'right' hypothesis at some stage. A more formal tree-based approach makes sure nothing is missed. For reasons such as these, a great deal of effort has been expended on developing tree classifiers in medicine.

Examples of medical classification trees are given in Choi *et al.* (1991) on prognosis after head injury, Mabbett *et al.* (1980) for classifying jaundice patients, and Goldman *et al.* (1982) for classifying patients with acute chest pain.

4.2.3 Social sciences

Amongst the earliest work on formal statistical methods for deriving classification trees was that carried out by social science researchers. The primary aim of this work seems to have been as an aid to *understanding* data rather than production of a tree which could be used to classify future objects, as illustrated by the names of the programs developed: *automatic interaction detection* (AID) (Morgan and Sonquist, 1963; Sonquist, 1970); *theta-AID* (THAID) where the prefix describes the splitting criterion (see below) of maximising the sum of the number of observations in each modal category (Morgan and Messenger, 1973; Fielding, 1977), and chi-squared AID (CHAID) which includes a statistical testing procedure based on the chi-squared statistic (Kass, 1980).

4.3 PARTITIONING CRITERIA

In a binary tree each nonterminal node is split into two branches. That is, the values of the variable defining the split at such a node are partitioned into just two classes, as in the two hypothetical examples above. Binary trees are the most common form. The restriction to just two branches is not severe since ordered trees with more than two branches can be mapped to equivalent binary trees. If the variable is an ordered one, the simplest and by far the most common way of doing this is to define a threshold such that objects with values less than the threshold go down one branch and objects with values greater than the threshold go down the other branch. If the variable is not ordered, an explicit partition is needed to assign each of the variable's categories to a particular branch.

When constructing a tree, we have to decide whether a particular, potentially non terminal, node should be split, and if so, on what variable it should be split and the nature of that split.

One way to resolve these problems is to use an (im)*purity index*—a measure, for a given node, of the differences between the probabilities of belonging to each class. For example, suppose a node contains 10 design set objects, 5 from class 0 and 5 from class 1, and suppose that all the class 0 objects have scores less than 6 on some variable x whereas all the class 1 objects have scores greater than 6. Then this node itself is relatively impure: it has equal numbers of design set objects from each of the two classes. A new object percolating down the tree and reaching this node might reasonably be estimated to have a 0.5 chance of belonging to each of the two classes. However, if we were to take the tree further, splitting this node using x at the value 6, each of the two offspring nodes would be perfectly pure. One would have only class 0 design set objects within it and the other would have only class 1 design set objects.

To generalise, consider a proposed split s of a node v, into two offspring nodes l and r. Let $i(v)$ be some measure of impurity of node v (with large values meaning impure), and let $\pi(l)$ and $\pi(r)$ be the proportions of the design set cases in node v which the split s puts into nodes l and r respectively. Then a measure of the change in impurity which would be produced by split s of node v is given by

$$\Delta i(s, v) = i(v) - \pi(l)i(l) - \pi(r)i(r)$$

Δi can now be used as a partitioning criterion: a high value means a proposed split is a good one. This allows us to make all three of the construction decisions. We can decide whether to split by examining if a proposed split will lead to an increase in purity. We can decide the variable to use and the nature of the split by examining all of the the variables and all potential splits on them to find the one which yields the greatest increase in purity.

Of course, these are not complete answers. The change in impurity index defined above is based on the design sample and hence is subject to sampling variability. We shall need to consider this; an apparent increase in purity may all too easily be attributable to sampling variation if the number of objects in node v is small. Similarly, it is all very well to say that we can choose where and how to split by examining all of the variables and all potential splits but this may be a large computational exercise. We shall also see that defining partitioning criteria via impurity indices is not always an ideal solution.

Once a measure of impurity has been chosen, in addition to testing the merits of particular splits, it can also be used to define a global measure of a tree's impurity:

$$I = \sum_{v \in T} i(v)\theta(v)$$

where T is the set of terminal nodes of the tree and $\theta(v)$ is the proportion of design set objects which fall in terminal node θ. This is thus a *resubstitution estimate* (see Chapter 7) of a particular measure of quality of the tree in question. Moreover, consider a particular tree A, and a proposed new tree A', formed by a split s of a terminal node t of A. Then

$$I(A) = \sum_{v \in T_A} i(v)\theta(v)$$

and

$$I(A') = \sum_{v \in T_A - t} i(v)\theta(v) + i(l)\theta(l) + i(r)\theta(r)$$

where l and r represent, respectively, the left and right offspring nodes of node t. Since $\theta(l) = \theta(t)\pi(l)$ and $\theta(r) = \theta(t)\pi(r)$ we have

$$\begin{aligned}
\Delta(I) &= I(A) - I(A') \\
&= i(t)\theta(t) - i(l)\theta(t)\pi(l) - i(r)\theta(t)\pi(r) \qquad (4.1) \\
&= \Delta i(s, t)\theta(t)
\end{aligned}$$

That is, the overall reduction in impurity of a tree caused by splitting a node t is proportional to the impurity reduction of that node, where the proportionality factor is the size of the node.

How might one choose an impurity index on which to base a partitioning criterion? Some such measures which have been used in the context of tree classifiers are as follows.

Index 1

An obvious measure of impurity of a node v is the (resubstitution) estimate of the probability that objects at this node belong to the class other than that with the maximum estimated probability. This is the resubstitution estimate of error rate: $i(v) = \{1 - \max_j \hat{p}(j|v)\}$. From equation (4.1), for a given node v, this will lead to choosing the split s which maximises

$$\Delta i(s, v) = \{1 - \max_j \hat{p}(j|v)\} - \{1 - \max_j \hat{p}(j|l)\}\pi(l) - \{1 - \max_j \hat{p}(j|r)\}\pi(r)$$

$$= -\max_j \hat{p}(j|v) + \max_j \hat{p}(j|l)\pi(l) + \max_j \hat{p}(j|r)\pi(r) \qquad (4.2)$$

Although superficially attractive, this measure has weakenesses. First, since max is a nonlinear function, we have in general that $\max(a_j) + \max(b_j) \geqslant \max(a_j + b_j)$, from which it follows that $\Delta i(s, v) \geqslant 0$ for all splits. In the special case of two classes it can occur that node v has a majority of class 0 objects and that, for all splits, both offspring nodes also have a majority of class 0 objects. From (4.2) it follows that $\Delta i(s, v) = 0$ for all splits. This means that one cannot identify a 'best' split.

A second drawback of this criterion is illustrated by Breiman *et al.* (1984) using the following example. Consider a node v containing 400 objects from each of two classes, denoted $v(400, 400)$ and let a split partition this as $l(300, 100)$ (meaning that the left-hand offspring node has 300 objects in class 0 and 100 in class 1) and $r(100, 300)$. Then, assigning classes to leaf nodes on a majority vote basis, this tree will misclassify 200 objects. Now consider the alternative partition into $l(200, 400)$ and $r(200, 0)$. This tree also misclassifies 200 objects. This tree, however, seems more attractive. In terms of probability estimates, it is very confident about 200 of its classifications, whereas the first tree is never very confident (see Chapter 6 for a discussion of the confidence one may place in classifications in general). Moreover, the second tree already has a leaf node— node r needs no further partitioning. What is needed is an impurity index which gives greater rewards to purer nodes.

Index 2

For the two-class case, we suggest in Chapter 6 that the variance of the probability distribution of belonging to class 0 might be suitable: $p(0|v)p(1|v)$, estimated by $\hat{p}(0|v)\hat{p}(1|v)$. This has several desirable properties (see Section 6.7): (1) it takes its maximum value of 1/4 when $\hat{p}(0|v) = \hat{p}(1|v) = 1/2$; (2) it takes its minimum when $\hat{p}(0|v)$ or $\hat{p}(1|v)$ is 1; and (3) it is a symmetric function of

$\hat{p}(0|v)$ and $\hat{p}(1|v)$. But does it give greater rewards to purer nodes than the resubstitution error rate of Index 1?

For the two-class case, we can write impurity index 1 as

$$i(v) = \{1 - \max_j \hat{p}(j|v)\} = \min_j \hat{p}(j|v) = \left\{ \begin{array}{ll} \hat{p}(1|v) & \hat{p}(1|v) \leqslant \dfrac{1}{2} \\[2mm] 1 - \hat{p}(1|v) & \hat{p}(1|v) > \dfrac{1}{2} \end{array} \right\}$$

That is, this index is a linear function of $\hat{p}(1|v)$ on either side of 1/2. To give greater rewards to purer nodes, we need to replace it by a concave function of $\hat{p}(1|v)$; this would mean that, for a given change in $\hat{p}(1|v)$, the function would give a greater change in purity than Index 1. More formally, we require that the function's second derivative should be negative for all $\hat{p}(1|v)$. Well, $\hat{p}(0|v)\hat{p}(1|v)$ $= \hat{p}(1|v) - \hat{p}(1|v)^2$, which does indeed have a negative second derivative. Thus $\hat{p}(0|v)\hat{p}(1|v)$ is a satisfactory measure of impurity for the two-class case.

How might we extend this to the multiclass case? An obvious extension is to $\sum_j \hat{p}(j|v)[1 - \hat{p}(j|v)]$. This is easily seen to be an estimate of the probability that a randomly chosed object from node v will be assigned to an incorrect class. It is perhaps more commonly written in the equivalent forms $i(v) = \sum_{j \neq k} \hat{p}(j|v)\hat{p}(k|v) = 1 - \sum_j \hat{p}(j|v)^2$, and is known as the *Gini index*.

Index 3

Another index which is sometimes used is defined as $i(v) = -\sum_{j=0}^{C} \hat{p}(j|v)\ln \hat{p}(j|v)$ where $\hat{p}(j|v)$ is the proportion of design set objects in node v which belong to class j. This satisfies the above criteria and is a natural index in the context of information theory, maximum likelihood and entropy.

Index 4

Automatic interaction detection (AID), one of the earliest programs implementing this kind of approach to classification (though with an emphasis on discerning the data structure, rather than building trees for classifying future cases) assumed an interval scale dependent variable and maximised the between-group sum of squares at each node. THAID handled nominal dependent variables—the sort of problem we are concerned with—and based a splitting criterion on the sum of the number of objects in the largest category in each offspring set.

Purity indices are all very well, but they are not the only way to think about partitioning and they may not be ideal. For example, even the partitioning criterion derived from the popular Gini index has its weaknesses. First, it has a tendency to produce offspring nodes that are of equal size. To see why this should occur, note that the measure of change in impurity derived from the Gini index can be written in the form

$$\Delta i(s,\ v) = \pi(l)\pi(r)\left[\sum_j \hat{p}(j|l)^2 + \sum_j \hat{p}(j|r)^2 - 2\sum_j \hat{p}(j|l)\hat{p}(j|r)\right]$$

The first factor in this expression is concave down and takes its maximum when $\pi(l) = \pi(r)$, so favouring equal splits. Equal splits may not be the most effective. Whether or not they are, the tendency to produce equal splits can be increased or decreased by modifying the criterion using extra factors involving $\pi(l)\pi(r)$.

Secondly, although a greater (Gini) purity for two-class problems means offspring nodes with majorities from different classes, this may not be the case with more than two classes. In response to this second disadvantage of the splitting criterion derived from the Gini index, Taylor and Silverman (1993) suggested the following partitioning criterion, which they call the *mean posterior improvement (MPI) criterion*:

$$\Delta i(s,\ v) = \pi(l)\pi(r) - \sum_j \hat{p}(j|v)\pi(l|j)\pi(r|j)$$

where $\pi(l|j)$ is the proportion of class j objects from node v which go to offspring l. This criterion is always positive. The MPI places greater emphasis on larger classes in node v and will be large when all objects from class j fall in the same offspring, for all j. It can equivalently be written as

$$\Delta i(s,\ v) = \pi(l)\pi(r)\left[1 - \sum_j \frac{\hat{p}(j|l)\hat{p}(j|r)}{\hat{p}(j|v)}\right]$$

(the ratio on the right is taken to be 0 if $\hat{p}(j|v) = 0$). From which it can be seen that, like the Gini index, the MPI tends to favour equal-sized splits.

The MPI also has the property of *exclusive preference*, defined by the following two conditions:

(i) Given $\pi(l)\pi(r)$, it takes its maximum value when $\sum_j \hat{p}(j|l)\hat{p}(j|r)$ is zero. This means that no class has objects falling in each offspring node.
(ii) Given $\pi(l)\pi(r)$, it takes its minimum value when $\hat{p}(j|l) = \hat{p}(j|r)$ for all j. That is, it is minimum when the split is ineffective.

Taylor and Silverman (1993) give an example showing that the Gini index does not satisfy exclusive preference.

Another node-splitting method, which is effective for problems involving many classes, is the *twoing* approach (Breiman *et al.*, 1984). At each node the classes are grouped into two superclasses. The optimum split is computed as if it were a two-class problem, and the grouping is found which yields the best split. A merit of this strategy is that the (possibly many) classes are divided into a hierarchy of subsets which are most similar, so aiding understanding and interpretation. Breiman *et al.* show that this approach is equivalent to finding the split which maximises the criterion

$$i(v) = \frac{\pi(l)\pi(r)}{4} \left[\sum_j |\hat{p}(j|l) - \hat{p}(j|r)| \right]^2$$

Alternatives to applying general measures of purity to choose each split are (i) to focus on individual classes versus the rest or (ii) to focus on distinguishing between two classes, letting the others divide up in an uncontrolled manner, to be sorted out lower in the tree.

Despite the above discussion, it seems that, as with the precise choice of kernel shape in smoothing methods (Chapter 5), the performance of a tree is relatively insensitive to the splitting criterion.

4.4   GROWING, STOPPING, PRUNING AND AVERAGING

In principle, at least for continuous data (where no two objects have *exactly* the same value on any variable), one could continue partitioning nodes until all the leaf nodes contained only a single object. This would lead to a zero resubstitution error. However, it would obviously grossly overfit the training data and would be very unlikely to generalise well to new data. It leads to overlarge trees with many 'twigs' which merely model random variation in the design set arising from the sampling procedure, rather than modelling any real underlying structure of the populations. To overcome this, we need to grow a shorter tree, perhaps to stop the growth before it reaches this extreme (see Section 1.3).

Early work on formal tree-growing methods did, indeed, adopt this approach of *stopping* growth. For example, an obvious stopping rule is to stop when the maximum reduction in impurity is less than some threshold. A small threshold will lead to many small leaf nodes, whereas a large one will lead to few large leaf nodes. In terms of the bias/variance trade-off discussed in Section 1.3, a small threshold has high variance with low bias, whereas a large threshold has low variance but high bias in the probability estimates. Such an approach has weaknesses, however. One of its problems arises because the basic approach to tree growth is *sequential*. A split will only be made if the impurity reduction at the proposed node exceeds some value, regardless of what might happen lower in the tree as a consequence of such a split. For example, the measurement space representation in Figure 4.3 shows that splitting variable x at 1 will lead to only a very slight decrease in purity—only a very small threshold value would lead to such a split being made. (And there is no split on y which leads to any decrease.) However, if the split at $x = 1$ was made then subsequent splits at $y = 1$ in both nodes would lead to a very good separation between the classes. The point is that, if a stopping criterion is used, the huge potential of the split on y following the split on x is unlikely to be discovered.

There are several alternatives to simply stopping tree growth. The most popular is to grow a large tree (small threshold; at an extreme, all nodes contain-

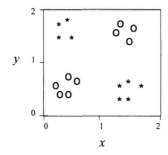

Figure 4.3 An example in which a split yielding a slight improvement in purity leads to a split yielding a dramatic improvement.

ing design set members belonging to only one class or nodes which cannot be made more pure by any split) and then *prune* it back. That is, leaf nodes are cut off, replacing them by their common parent. This smooths out the overcomplex decision surface. Of course, to implement this, one needs to decide when to prune and when not to prune a pair of leaves. Growing and pruning trees are analogous processes to forwards and backwards stepwise regression (Section 9.1). In the former one sequentially adds the regressor which minimises the predictive accuracy (typically R^2) and in the latter starting from the saturated model containing all regressors one sequentially eliminates them on the basis of which one contributes least to the predictive power of the model.

One obvious criterion to decide which leaves to prune is to use $\max_j \hat{p}(j|v)\theta(v)$ as a measure of the value of each leaf node. Objects in such a node are assigned to the class maximising $\hat{p}(j|v)$, so that the first factor is an estimate of the probability that points at this node will be correctly classified. The second factor serves as a weight indicating the size (importance) of the node. A large value of this criterion means that not only does it seem correctly to classify most objects which arrive at it (according to the resubstitution estimate), but also that large numbers of points do actually arrive. A small value means that either the node is small (so it does not contribute greatly to the tree's performance) or that the probability estimates at that node are not strongly in favour of the largest class, or both. Overall, the pruning to be adopted will have the smallest criterion value.

Just as backwards stepwise regression can be generalised so that more than one variable is removed at each step—e.g. remove that *pair* whose removal leads to least degradation in performance—so can this process be generalised by considering the removal of more than a single pair of leaves at a time.

As is stressed elsewhere in this book, the conditions of a problem will determine what criterion should be optimised by the design of the classifier. Rather than simply trying to optimise classification performance, one might also wish to take account of the size of the tree, on the principle that simpler trees are to be preferred, all other things being equal. Then the overall quality of a tree is measured by $\sum_{v \in t}\{1 - \max_j \hat{p}(j|v)\}\theta(v) + \alpha|T|$, where $|T|$ indicates the number of nodes in T and α is a parameter determining the payoff between predictive accuracy on the design set and complexity of the tree. In essence this is a rather ad hoc way of reducing the overfitting due to modelling the design set. By varying α, and for each value choosing the tree which minimises this criterion, a sequence of 'best' trees is produced. At one extreme (large α), minimisation of complexity becomes paramount, producing a very small simple tree—just the root node. At the other extreme, accuracy reigns, and the tree is large and complex. The final choice of tree from this sequence can be made on the basis of an error rate estimate, perhaps by using a test set or by cross-validation.

Number of nodes is one indicator of tree 'size'. Others include maximum path length and expected path length from the root to a leaf node. Clearly measures such as these are relevant when speed of classification is at a premium.

Apart from pruning, other approaches to finding a good tree structure include dynamic programming (perhaps an obvious technique in this application), branch and bound methods, and tree averaging. Tree averaging focuses on the type of model constructed rather than on the search algorithm; this makes it conceptually different, so we shall say something about it.

In tree averaging, instead of finding a single 'best' tree, one constructs several, even many, trees and averages their class predictions to produce a final classification. Clearly, in order to be able to do this, a key issue will be choosing the weights with which to average the different trees; Oliver and Hand (1996) adopt a minimum message length approach. In effect, this accords each tree a prior probability depending on its complexity: it is a practical application of Occam's razor, with simpler trees being deemed more likely.

The next question is what set of trees to average over. In principle, at least, one could average over all possible trees constructed from a given design set. In practice, however, for anything but a very small problem, the number of trees will be too large. Of course, choosing a subset can be regarded as averaging over all of them, but giving zero weights to those not in the subset. Several ways have been suggested for choosing the subset:

- Beginning with a complete tree, constructed to classify the new object in question, sequentially prune it back along the path leading to the object's final leaf node. The *path set* of trees is the set of trees obtained by this pruning process. Averaging is over the path set.

- The *option set* of trees is the union of path sets obtained from pruning back the complete set of trees. Clearly this is much more computationally demanding than merely using a single path set, so it may not always be practicable.

- To construct the *fanned set* of trees, take the path set and expand it at each leaf node, along the path of the object to be classified, by every possible split. This is intermediate between using merely the path set and using the option set.

4.5   ASSIGNING CLASS LABELS TO TERMINAL NODES

The most obvious way of assigning class labels to terminal nodes is via a simple majority vote. That is, assign all points which reach terminal node v to class k if $\hat{p}(k|v) = \max_j \hat{p}(j|v)$; if more than one class takes the maximum then assign randomly to one of these classes. This will need to be adjusted if the sample sizes in the design set are different from the class priors or if different misclassification costs are involved. This approach generalises immediately to yield vectors of the estimated probabilities of belonging to each class, for any given node.

4.6   GRAPHS, TRELLISES AND BEYOND

We noted above that orthogonal partitions may not always be ideal. A simple bivariate example is given in Figure 4.4. Here we can see immediately that a diagonal split would separate the two classes very simply and easily, but that orthogonal splits lead to a large and awkward stepwise partition which in no way elucidates the underlying structure. Figure 4.4 also indicates the way to over-come the problem: generalise the concept of split to include thresholds on linear combinations of variables.

Although this generalisation leads to a more flexible class of trees, it is not without its disadvantages. In particular, the search space (which was already large) can now be astronomical. Moreover, it does not necessarily lead to improved interpretation: a sequence of splits on linear combinations can be complex to explain. Some simplification can be achieved by using only a subset

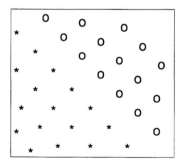

Figure 4.4: An example in which orthogonal splits will lead to a complex tree, but a single diagonal split will perfectly separate the two classes.

of variables in each linear combination; this subset is found by stepwise methods. Once a suitable subset has been identified, the decision surface in the space spanned by that subset has to be found (the two processes usually run hand in hand). That means a suitable linear combination has to be found, effectively reducing the subset to a single dimension, before a threshold is imposed. The scope here is clearly vast, since any standard method of formulating a linear classifier can be used. Loh and Vanichesetakul (1988) use standard linear discriminant analysis at each node.

The use of linear combinations in trees has its converse in the use of trees in linear combinations. Such a strategy is adopted for credit scoring, where the basic models are linear and any interactions are introduced through new variables constructed from small trees of measured variables (called derogatory trees).

We have noted elsewhere that in the machine learning community it is perhaps more common to visualise things in terms of categorical variables, especially binary, rather than continuous variables. Linear combinations of categorical variables are not such an obvious way of overcoming interaction problems and, instead, new variables are defined as Boolean combinations of the raw ones. Again there are many possible derived variables and a subset is needed to make the method practicable.

Trees have the property that proceeding down a branch is irrevocable: once a split has occurred, the objects proceeding down each arm of the split are necessarily treated as different groups. That this can be inefficient is perhaps most readily seen from the machine learning perspective.

Suppose we have four binary attributes a, b, c and d, and the (unknown) optimal division is into the set S satisfying [(a and b) or (c and d)] and its complement. Suppose we begin the tree by testing a and b. If both are true then the object is a member of S. However, if either a is false or a is true but b is false, then it is necessary to see if (c and d) is true. That means there are two branches of the tree, one for a false and one for a true and b false, for which we need a small subtree testing (c and d). The subtree is identical in these two cases. This is an example of the *replication problem*. From a more statistical perspective, the subtrees, supposedly identical, will be learnt from different subsets of the training data, with consequent inaccuracy due to sampling variation.

A similar problem with continuous variables was overcome by permitting a partition at a node to depend on a linear combination of variables, rather than a single variable. An analogous strategy here is to replace the single variable at each node by a Boolean combination of variables. Of course, this process is merely feature definition. An alternative strategy is to allow the merging of different branches of the tree. This means that the structure is no longer a tree, but becomes a more general directed graph, a *decision graph or trellis*. In the above example, the branch from a false and the branch from a true and b false would both lead to the same subgraph, the subgraph testing (c and d).

This sort of strategy seems more natural with nominal variables (as in the example) than with interval or continuous variables, where the ordinal nature imposes an implicit continuity of class membership probabilities.

4.7   OTHER TOPICS

In terms of a measurement space representation, missing values in objects to be classified require decisions to be made in marginal spaces. For some types of classification rule this poses no problem, but this is not the case with trees. As an object percolates down the tree, difficulties will occur when it reaches a node requiring a decision to be based on a variable which has not been measured. One way to tackle this is to define *surrogate splits*, splits which most accurately predict the effect of other splits.

If two variables have a very similar effect on the probabilities of class membership, then (a) one may not appear in a tree, even though it is strongly related to the probabilities, and (b) the structure of the tree may be unstable in the sense that a slight change in the design set could lead to a substantial change in the tree topology. Point (a) is familiar with other classifiers in the context of multi-collinearity. Point (b) is a manifestation of the *flat maximum* effect discussed in Section 9.3. When the aim is classification, point (b) will not matter, but it will matter when the aim is understanding.

Although by far the greater part of the work on tree classifiers has used the general *top-down* approach outlined above, it is also possible to adopt a *bottom-up* approach, similar to hierarchical clustering algorithms in which nearest points and clusters are sequentially merged; see, for example, Landeweerd *et al.* (1983).

4.8   FURTHER READING

As noted above, work on recursive partitioning methods is ubiquitous. This makes it impossible to give a complete list of papers describing or adopting such methods. Nonetheless, one can identify important and influential publications.

The paper by Payne and Preece (1980) presents a comprehensive review of early formal work in the area. Although it appeared in a statistical journal, it approaches things from the background of diagnostic keys in biological applications and this may explain why it has not received the prominence it deserves. For example, the influential book by Breiman *et al.* (1984), which introduced the methodology underlying the CART program, did not include it in its reference list. Prior to this book, the statistical papers on the topic had tended to appear in isolation and the methods were regarded as rather offbeat (they were so different from linear models, for example). The mathematical rigour of Breiman *et al.* and the unification it yielded, lent these methods a new respectability. The book formalised the procedures, whereas before they had been seen as largely ad hoc.

One of the early influences on the machine learning community was Hunt's *Concept Learning System* (Hunt *et al.*, 1966). For example, the approach described in Quinlan (1982) is based on this. The Concept Learning System used cost as its basis for selection of variables, but Quinlan adopted a measure of information, more in keeping with most modern approaches.

The blossoming of parallel work on essentially the same problem in the two areas of statistics and machine learning soon led them to come together; see, for example, the proceedings of the biennial conferences on artificial intelligence and statistics, Hand (1990), Hand (1993), Cheeseman and Oldford (1994). The paper by Payne and Preece (1980) is particularly interesting in this context because it covers aspects showing the differences in emphasis of each of the machine learning and statistical communities. For example, of special interest to machine learning theorists, it describes the notions of *redundant* and *irredundant* sets of characters, a concept implicitly based on pure separability, and of optimal identification keys. It also includes discussion of probabilistic and decision theoretic methods.

Alternately extending and contracting a model—in this context growing and pruning a tree—is a standard strategy for searching through a potentially vast model space for a well-fitting model. In particular, it has a long history in the statistical pattern recognition context (e.g. Batchelor, 1968). An alternative to deciding what to prune using a measure based on the design data has been proposed by Gelfand *et al.* (1991), who suggested splitting the data into two sets, growing on one and pruning on the other.

In the introduction, we described how machine learning approaches sometimes transformed trees into rule bases. This is not the only way of formulating rule bases. An alternative approach is to begin with a single 'seed' rule for assigning objects to a particular class, to modify that rule by trying various changes to its predicates, and to discard any modifications which lead to misclassifications (note the machine learning emphasis on total accuracy, in contrast to the statistical tolerance of errors). Conjunction of the results of this with the results derived from other seed rules forms the complete classifier. This is in essence a bottom-up strategy which can be extended in various ways. Examples of work on machine learning of rules are Clarke and Niblett (1989) and Quinlan (1993).

Trees provide a natural and easily grasped way of visualising a hierarchical classification process. In general, if interpretation and understanding is important, one will probably want to split each node on a single variable, rather than a combination of variables. But this will not matter if classification performance is the objective regardless of comprehensibility.

The simple display of a tree as a series of connected nodes has been extended by Taylor and Silverman (1993) who display *block diagrams*. In each node a plot of the objects shows the values they take on the variable which will be used to partition that node. This gives a visual representation of the classes' separability.

Other criteria for splitting include methods based on the Kolmogorov–Smirnov statistic (Rounds, 1980), based on the principle that this identifies where two probability distributions take equal values, precisely the place at which we want to place the decision surface; the chi-squared statistic (Sturt, 1981), identifying the split leading to nodes for which the distributions of objects across classes are most different, so this can be applied with any number of classes; permutation statistics (Li and Dubes, 1986), again measuring the difference between distributions; the between-class to within-class variance ratio (Loh and Vanichesetakul, 1988); information gain (Quinlan, 1986), and minimum description and minimum message length (Quinlan and Rivest, 1989; Wallace and Patrick, 1993). Mingers (1989) studies some of these measures.

Tree averaging has been described by Buntine (1992), Hastie and Pregibon (1990) and Quinlan (1992). Decision graphs and related structures have been explored by Chou (1988, 1991), Mahoney and Mooney (1991), Oliver (1993) and others. Further reviews of decision tree classifiers are given in Safavian and Landgrebe (1991) and Michie et al. (1994, Chapter 5).

Nonparametric Smoothing Methods

5.1 INTRODUCTION

One way of looking at classification rules is to regard them as comparing estimates of the probabilities of belonging to each class, $f(j|\mathbf{x})$ for $j = 1, ..., c$, with each other, allocating an object with measurement vector $\mathbf{x}$ to the class with the largest $\hat{f}(j|\mathbf{x})$ (perhaps weighted to take costs into account). In order to obtain such estimates, one can go directly for the posterior probabilities $f(j|\mathbf{x})$, or one can estimate them indirectly via the class conditional distributions $f(\mathbf{x}|j)$ using Bayes' theorem: $f(j|\mathbf{x}) = f(\mathbf{x}|j)\pi_j / \sum_k f(\mathbf{x}|k)\pi_k$. In the Bayes' theorem approach, since the denominator is the same for all j, it can generally be ignored in the comparisons.

In this chapter we examine two nonparametric methods for estimating the relevant probabilities. The *kernel* method is usually described in terms of producing estimates of the class conditional distributions, transforming them as above to yield estimates of the $f(j|\mathbf{x})$. The *nearest neighbour* method is usually thought of as providing direct estimates of the $f(j|\mathbf{x})$.

5.2 THE KERNEL METHOD

The most widely used nonparametric approach to estimating probability functions is the histogram. For a univariate continuous variable, this is defined by partitioning the range of possible values of the variable into several cells of equal length and plotting for each cell a bar with length proportional to the number of observations falling in that cell. Such estimators could be used (and, indeed, sometimes are) to yield a classifier: a new point with measurement x would be classified by comparing the heights of the histograms at x for each class. Of course, if the samples for each class had only a few cases, so that the heights of the histogram bars could take only a few values, then there would be a high probability that the estimated probabilities $\hat{f}(\mathbf{x}|j)$ (proportional to the heights of the bars) would be equal. This is one potential disadvantage.

But there are also more serious disadvantages. In particular, the histogram will have discontinuities in its probability estimates at each cell boundary. This seems unnatural, with continuous variables anyway, where one usually expects some kind of smooth variation. Also, to produce a histogram, it is necessary to decide how wide the cells are, how many there are, and where the cell boundaries should be located (these three properties are related, of course).

In essence, the probability estimate within a histogram cell is an estimate of the average of the probabilities for x values in that cell. If the cell is large, so that different x values may correspond to very different probabilities, the histogram estimate may be very biased for some parts of the cell. On the other hand, large cells will tend to have more observations falling within them, so the corresponding probability estimates will have small variance. Conversely, a small cell inevitably means that few observations will lie within it, so the variance of the probability estimate for that cell will be large; but small cells, covering only a small range of x, have the advantage of small bias. These issues are merely an elementary manifestation of the bias/variance trade-off discussed in Section 1.3. Clearly, to choose 'the best' cell width one really needs to adopt some overall performance criterion (such as mean squared error summed over cells) and choose the cell width and location to minimise this. But this does not overcome the problem of the discontinuities.

A further issue which must be addressed is the extension to more than one variable. The obvious extension of the histogram is to partition the space according to the product of the partitions of each variable. However, this falls foul of the *curse of dimensionality*. If each variable is separately divided into 10 cells, and if there are 10 variables, then there are 10^{10} cells in the multivariate space. Few samples are of a large enough size to produce accurate probability estimates in so many cells! Indeed, most of the multivariate cells will be empty, so classifications may be based on comparing probability estimates of zero with probability estimates of zero. Not the most useful of classification rules.

Kernel density estimation is a generalisation of the histogram approach which overcomes the discontinuity problem, the problem of where to locate the cells, and to some extent, the multivariate nature of most classification problems. There remains the problem of how wide the 'cells' should be and various estimates have been suggested.

To introduce kernel approaches we shall start with a single variable and, for notational simplicity, we shall temporarily drop the reference to the particular class involved; the kernel density estimation method is applied separately to each class, then these estimates are combined using Bayes' theorem. We shall therefore represent $f(x|j)$ by $f(x)$. Also, we shall develop the ideas assuming continuous variables, extending them to the categorical case later.

Our aim, then, is to estimate the probability density function, $f(x)$, at x from a sample of n points randomly drawn from $f(x)$. Clearly, the greater the proportion of the design sample points lying within the vicinity of x, the greater should be our estimate. So, an obvious approach would be to take an interval, length h say, centred on x, and estimate the probability $f(x)$ by the proportion of the n points which fall within this interval. Put another way, each point within h of x contributes $1/n$ to the estimate at x and each point further away than h contributes nothing. The estimated probability at x is thus a weighted sum of contributions from the n points. Of course, it is a rather crude weighted sum, and one may object to it on these grounds: why, after all, should points within h contribute the full amount $1/n$ to the estimate and points just beyond h contribute nothing? This is almost as bad as a rigid adoption of a 5% level in significance testing! What it means is that, as we move x along the line, whenever a design set point enters or leaves the $[x - h, x + h]$ interval, there will be a discontinuous jump of $+1/n$ or $-1/n$. We can overcome this by replacing the crude 0 or $1/n$ weights by a smoother set of weights. In particular, we can let the size of the weights decay as the distance between x and the design set point in question increases. We could, for example, use a Gaussian curve as the weight function.

This approach has overcome the histogram problems of where to locate the cells (by positioning a cell at x, wherever that is) and of the discontinuities. It remains to decide on the precise form of the (now assumed smooth) weight decay function. These functions are normally called *kernel* functions. This form has two aspects. The first is its actual shape. Should it, for example, be the sharp cutoff of our first example, or should it be the smooth bell-shaped curve of the second example, or should it be something else? The second aspect is the width of the kernel function; its *spread, bandwidth, window width* or *smoothing parameter*. This is h in the first example and will be the standard deviation of the Gaussian curve in the second example. If h is too small the estimate will be highly irregular (tending towards the unsmoothed empirical distribution as $h \to 0$). Conversely, if h is too large the estimate will be too smooth, tending towards a constant value as $h \to \infty$.

We now express the above more formally. Denoting the kernel function by $K(.)$, the estimate of the probability density at x, using design sample $\{X_i\}$, is

$$\hat{f}(x) = \frac{1}{nh} \Sigma_i K \left(\frac{x - X_i}{h} \right)$$

If K is itself taken to be a unimodal probability density function then $\hat{f}(x)$ will itself be a density. This will ensure that $\int \hat{f}(x)dx = 1$ and that $\hat{f}(x) \geq 0$ for all x.

Our description of the kernel method above focused on the point x and described each design set point as contributing an amount depending on the ordinate, at X_i, of the kernel function centred at x. An equivalent way of looking at it is to imagine the $1/n$ contribution from each design set point being spread out according to the shape of the kernel function. The sum of the ordinates, at x, of these spread functions yields the estimate. The overall resulting estimate is the convolution of the empirical distribution function with the kernel function.

To use the kernel method we need to specify the kernel shape and the size of the smoothing parameter. To do this—to make good choices for these two aspects of the estimator—we first need to decide what we mean by 'good'. In the context of kernel density estimation, a common choice is the asymptotic mean integrated squared error. We begin the discussion with the smoothing parameter.

5.2.1 Choice of smoothing parameter

The choice of the smoothing parameter h determines the trade-off between bias and variance at each x. A common measure combining these two aspects of accuracy, and so yielding a single measure which can be optimised by choice of h, is mean squared error (MSE). Ideally, when a particular object with a particular x value is to be classified, we will want to know the best value of h for *that* x. However, to produce a classifier which can be used for *any* future object without going to the trouble of finding the corresponding optimal h each time, we would like to find some 'overall optimal' h. To do this, we need to combine the MSEs at each x into a global measure. An obvious way is to integrate over the entire space of x. That is, we can define a global criterion indicating how close $\hat{f}(x)$ is to $f(x)$ by $\int E\,[\hat{f}(x) - f(x)\,]^2 dx$. By changing the order of the two integrals involved, this is mean integrated square error (MISE).

Of course, what we are talking about here is a measure of the difference between two probability density functions, so any of the measures outlined in Chapter 6 could be used. Moreover, MISE is certainly not the criterion by which the final performance of the classification rule will be judged, so it is arguable that this is not the most appropriate criterion to use (see Section 6.1). However, MISE seems to be the criterion most frequently adopted in kernel smoothing.

We have

$$MISE = \frac{1}{n}\int\left\{\int \frac{1}{h^2}K^2\left(\frac{x-y}{h}\right)f(y)dy - \left[\int \frac{1}{h}K\left(\frac{x-y}{h}\right)f(y)dy\right]^2\right\}dx$$

$$+ \int\left\{\int \frac{1}{h}K\left(\frac{x-y}{h}\right)f(y)dy - f(x)\right\}^2$$

which is a complicated function of h. An asymptotic approximation to this is given by

$$AMISE = \frac{1}{nh}\int K(x)^2 dx + \frac{1}{4}h^4\left[\int x^2 K(x)dx\right]^2 \int f''(x)^2 dx$$

a much more straightforward function of h. Using this, we can obtain the optimal h as

$$h_{OPT} = \left[\frac{\int K(x)^2 dx}{n\{\int x^2 K(x)dx\}^2 \int f''(x)^2 dx}\right]^{1/5} \tag{5.1}$$

Equation (5.1) shows that, when f is irregular so that $\int f''(x)^2 dx$ is large, h_{OPT} will be small. This is the sort of behaviour we hope to see. Of course, we cannot use (5.1) directly since it makes use of the unknown f. However, we can use it as the basis for other methods.

If one has some idea of the general sort of shape of f then a simple approach is to use h_{OPT} assuming that form. For example, assuming f to be normal with standard deviation estimated by $\hat{\sigma}$ leads to

$$h = \hat{\sigma} \left[\frac{8\pi^{1/2} \int K(x)^2 dx}{3n\{ \int x^2 K(x) dx \}^2} \right]^{1/5}$$

Also, one can use a kernel estimate of $f''(x)$ in (5.1) to lead to an iterative approach.

Criteria based on MSE are all very well, but a deep justification for them may be difficult to find, perhaps especially in a classification context. Error rate is obviously a more justifiable criterion in this context, i.e. one chooses the h for each class, or the $\mathbf{H}$ matrices in the multivariate case (see below) to minimise the error rate. But error rate leads to difficult optimisation issues. Likelihood is another general criterion, and is generally sufficiently tractable to be used. However, difficulties arise in the context of kernel estimation. The likelihood function, evaluated on the design set $\{\mathbf{X}_i\}$, will involve terms like $K((X_i - X_i)/h)$, which will lead to the maximum likelihood occurring at $h = 0$.

A common way to sidestep this problem is the *leaving-one-out* approach, evaluating the likelihood at X_i of the estimator based on all design set points except X_i. That is, one chooses the h which maximises

$$\prod_i \hat{f}_{/i}(X_i) = \prod_i \sum_{\substack{j \\ j \neq i}} \frac{1}{hn} K\left(\frac{X_i - X_j}{h} \right)$$

5.2.2 The kernel shape

One of the problems of deciding what kernel shape to use is that the shape and smoothing parameter are linked; we see above that the AMISE involves two terms, both of which include terms in h and terms in K as factors. This means that choosing the best K will also involve h. For comparing kernel shapes, we could try adopting equal bandwidths, but the essential arbitrariness in the shapes makes this meaningless. For example, we can think of

$$K_1(x) = \frac{1}{\sqrt{2\pi}} \exp(-x^2/2)$$

and

$$K_2(x) = \frac{1}{3\sqrt{2\pi}} \exp(-x^2/18)$$

as representing exactly the same curve but using different units of measurement. Choosing $h = 1$ in both of them will lead to the second curve having three times the dispersion of the first. As it happens, we can exploit the form of the AMISE to identify a particular member of each family of shapes; this is done as follows. Since the dispersion of K is arbitrary, we can replace K in all of the above by $K_\delta(x) = K(x/\delta)/\delta$. If we then choose

$$\delta = \left\{ \int K(x)^2 dx / \left[\int x^2 K(x) dx \right]^2 \right\}^{1/5}$$

it turns out that the AMISE factorises as

$$AMISE = \left\{ \left[\int K(x)^2 dx \right]^4 \left[\int x^2 K(x) dx \right]^2 \right\}^{1/5} \left\{ \frac{1}{nh} + \frac{1}{4} h^4 \int f''(x)^2 dx \right\} \quad (5.2)$$

That is, the kernel shape and the smoothing parameter only appear in separate factors. The particular member of the family with the value of δ which permits this is called the *canonical* kernel for that family. The particular kernel shape which minimises (5.2) is found by minimising the first factor. It turns out to be

$$K(x) = \left\{ \begin{array}{ll} 0 & |x| \geq 1 \\ \frac{3}{4}(1 - x^2) & |x| < 0 \end{array} \right\}$$

and is called the *Epanechnikov* kernel.

Although the Epanechnikov kernel is optimal in the sense indicated above, it turns out that other 'reasonable' choices of kernel shape do not lead to a very great loss of efficiency. For example, measuring efficiency of a kernel K as the ratio of the first factor of (5.2) for that kernel to the corresponding first factor for the Epanechnikov kernel, we find that Gaussian kernels have an efficiency of 95%. Also, if we relax the requirement that the kernel be itself a probability density function (for example, by allowing it to take negative values), it is possible to obtain better convergence with increasing n. In some applications of kernel density estimation, the consequence that the resulting estimate $\hat{f}(x)$ is not itself a probability density function will be seen as a disadvantage. In the present context, however, where future classifications will be specific to the x for the new object in question, it is not obvious that it is a serious disadvantage; local properties are more relevant than global properties.

One potential weakness of the basic kernel approach is that it adopts the same degree of smoothing for all x, regardless of how locally dense are the observation points or of how variable the underlying probability density function is thought to be in the vicinity of x. *Local* kernel approaches seek to overcome this by letting the size of the smoothing parameter h vary across the space. That is, h is allowed to be a function of x, the point at which the density estimate is to be made. Again this typically has the consequence that $\hat{f}(x)$ is no longer a probability density function. A complementary approach to overcoming the same weakness is to use *variable* kernel methods, in which the kernel function associated with each design set point is allowed to take a different h. In essence this means that the estimate is based on a sum of kernels with different widths.

5.2.3 Multivariate kernels

The univariate kernels discussed above are all very well, but they allow us to tackle rather limited classification problems. We need to generalise them to situations involving several, ideally many, variables. The most general multivariate form is

$$\hat{f}(\mathbf{x}) = \frac{1}{n}\sum_i K(\mathbf{x}, \mathbf{X}_i)$$

where $\mathbf{x}$ is the point at which the estimate is required and $\mathbf{X}_i$ is the ith design set element, all points being in d dimensions. As for the univariate case, we restrict consideration to K of the form $K(\mathbf{x} - \mathbf{X}_i)$. Further restrictions on the form of K are required to yield practical estimators, and the standard approach is to use ellipsoidal functions, typically probability density functions as in the univariate case. In this case, the extent of smoothing (corresponding to h in the univariate case) forms a *smoothing parameter matrix*. Two common forms are as follows:

(i) Multivariate Gaussian kernels yield the estimator

$$\hat{f}(\mathbf{x}) = \frac{1}{n(2\pi)^{d/2}|\mathbf{H}|^{1/2}}\sum_i \exp\left\{-\frac{1}{2}(\mathbf{x} - \mathbf{X}_i)'\mathbf{H}^{-1}(\mathbf{x}-\mathbf{X}_i)\right\}$$

with smoothing parameter matrix $\mathbf{H}$. This has the form of a mixture distribution, with a component placed at each sample point.

(ii) *Product kernels* are so called because they are formed as the product of univariate kernels:

$$\hat{f}(\mathbf{x}) = \frac{1}{nh_1...h_d}\sum_i \prod_j K_j\left(\frac{x_j - X_{ij}}{h_j}\right)$$

$\mathbf{H}$ is diagonal here. Other variants, generalising and restricting these, can easily be developed, but for all the usual reasons it is desirable to keep the number of parameters small (one could have a distinct multivariate Gaussian kernel for each design set point, each with its own arbitrary covariance matrix, but this would be of limited practical value). Typically, kernel (ii) is restricted so that its shape K_j is the same for all variables, $j = 1, ..., d$.

Much of the univariate theory relating to MISE carries over to the multivariate case. Although it is generally impossible to find explicit expressions for the optimal $\mathbf{H}$, as measured by AMISE, they can be obtained for kernel (ii) with $K_j = K$. We have

$$
h_{OPT} = \left[\frac{d \int K^2(\mathbf{x}) d\mathbf{x}}{n(\int x^2 K(\mathbf{x}) d\mathbf{x})^2 \int \left(\sum_{j=i}^{d} (\partial^2/\partial x_j^2) f(\mathbf{x}) \right)^2 d\mathbf{x}} \right]^{1/(d+4)}
$$

(Since $K_j = K$, $\int x^2 K(\mathbf{x}) d\mathbf{x}$ is independent of which variable, x, is chosen.) We see from this that things are more complicated from a practical perspective since multivariate integrals are required.

5.2.4   Categorical data

At the beginning of Section 5.2 we introduced the kernel method via a uniform kernel. This is the simplest form. Generalising these ideas to the multivariate case, the method estimates the probability at point $\mathbf{x}$, based on design set $\{\mathbf{X}_i\}$, using the kernel

$$
K(\mathbf{x}, \mathbf{X}_i) \propto \begin{cases} 1 & d(\mathbf{x}, \mathbf{X}_i) < t \\ 0 & d(\mathbf{x}, \mathbf{X}_i) \ge t \end{cases}
$$

where $d(\mathbf{x}, \mathbf{X})$ is a distance measure and t is a threshold. We can apply this immediately to the case of the variables being categorical. The disadvantages of this kernel shape when used with continuous data also apply with categorical data. Note that if t is less than the minimum distance between cells it produces the ordinary multinomial estimate. What we really need, in parallel with the continuous case, is a kernel which decays gradually with increasing distance from $\mathbf{x}$.

A suitable kernel for binary data is a product kernel where each factor has the form

$$
\lambda^{1-|x_j - X_{ij}|} (1 - \lambda)^{|x_j - X_{ij}|}
$$

Such a kernel has a factor λ for each variable where x_j and X_{ij} agree and a factor $(1-\lambda)$ for each variable where they disagree. That is, the factors in the product kernel have the form

$$
K_j(x_j, X_{ij}) = \begin{cases} \lambda & x_j = X_{ij} \\ 1 - \lambda & x_j \ne X_{ij} \end{cases}
$$

If λ is greater than $1/2$, the kernel shape decays as the distance between $\mathbf{x}$ and $\mathbf{X}_i$ increases. Here λ plays the same role as h in the continuous case. Also as in the continuous case, one can generalise this to allow different λ for each variable (as well as different λ for each class).

To extend this to the case of nominal (unordered categorical) variables in which categorical variable j has $g_j > 2$ levels, we can use the factors

$$K_j(x_j, X_{ij}) = \begin{cases} \lambda & x_j = X_{ij} \\ \dfrac{1 - \lambda}{g_j - 1} & x_j \neq X_{ij} \end{cases}$$

in the product kernel. And again, this can be generalised to allow different λ for each variable.

Intrinsic to the kernel method is a choice of distance measure. For continuous and binary variables with product kernels, the only issue in choosing this measure is how to combine different variables (that is, how to determine the relative importance of each of the variables). For more general nominal variables, scores on a particular variable are either the same or they are different; except perhaps for some unusual cases, there is no sense in which the sizes of differences can be compared. For ordinal variables, however, things are more difficult; the distances between different levels generally cannot be compared, let alone the distances between levels on different variables. The simplest solutions to this problem impose a metric on the relative distances.

In practice, the most popular approach to estimating the smoothing parameters λ is the leaving-one-out version of the maximum likelihood (ML) approach, as in the continuous case; again this is because, without leaving one out, the ML solution gives $\lambda = 1$. But this method can lead to occasional zero probability estimates.

5.3 NEAREST NEIGHBOUR METHODS

Kernel methods essentially focus attention on estimating the separate distributions of the classes (even if the smoothing parameters may be estimated by reference to some global classification performance criterion). In contrast, nearest neighbour methods go directly for the posterior probability of class membership. In fact, as will soon be apparent, the two types of method are closely related and the distinction becomes blurred at the interface between them.

Our objective is to estimate the conditional probability that an object with measurement vector $\mathbf{x}$ will belong to class j. An estimate of this probability is given by the proportion of design set points in the neighbourhood of $\mathbf{x}$ which belong to class j. In particular, if we define 'neighbourhood' by the distance from $\mathbf{x}$ to the kth nearest point from the design set, we can estimate the probability by the proportion of design set points which are in class j amongst the k nearest to $\mathbf{x}$. Using this proportion of the design set amongst the k nearest neighbours which belong to class j as the estimate of $f(j|\mathbf{x})$, classification is to the largest of the $\hat{f}(j|\mathbf{x})$ (subject to adjustment for costs as well as priors, if the design sample does not reflect this).

Here k serves the role of the smoothing parameter in kernel estimation: a larger k will mean less variance in the probability estimates, but is likely to introduce more bias in the estimate of $f(j|\mathbf{x})$. Conversely, a smaller k will mean greater variance and less bias. The difference between h and k is that the 'near' neighbourhood identified by the k varies in size according to the local density of design set points. Thus nearest neighbour methods are closely related to the 'local' kernel methods mentioned in Section 5.2.

The parameter k tells us how large a 'near neighbourhood' to consider, but it does not tell us what shape it is (relevant only in the multivariate case). That is, we still need to choose a metric by which to measure nearness. This is analogous to the problem of multivariate kernel shape discussed in Section 5.2.

A standard approach is to use the Euclidean metric $\{(\mathbf{x} - \mathbf{y})'(\mathbf{x} - \mathbf{y})\}^{1/2}$. By default, this treats the variables as if they were of equal importance (although it may be subject to an essentially arbitrary choice of measurement unit for non-commensurate variables). This seems difficult to justify; except on asymptotic grounds, which may have little relevance to real problems. Common standardisations for the variables (dividing by their standard deviations, for example) and attempts to make them of 'equal importance' impose a notion of relative importance of the variables which may have no relationship to their relative importance for discriminatory purposes.

Nearest neighbour methods essentially estimate the average probabilities over a local neighbourhood. If all the probabilities within that local neighbourhood are the same, the nearest neighbour approach will be an unbiased estimate of the probability. (More generally, if $f(j|\mathbf{x}) = \int_n f(j|\mathbf{X})f(\mathbf{X})d\mathbf{X}$ the method will be unbiased. Here the integation is over the k nearest neighbourhood region and $\mathbf{x}$ is the 'centre' of that region, the point at which an estimate is to be made.) Thus, if we chose the shape of the neighbourhood region so that it included only points $\mathbf{X}$ with $f(j|\mathbf{X}) = f(j|\mathbf{x})$, an unbiased estimate would result. Of course, we cannot do this since we do not know the $f(j|\mathbf{X})$ and $f(j|\mathbf{x})$. However, we could make an initial approximation to these probabilities then base a secondary estimate on that approximation. There are several ways this might be attempted.

One could initially approximate $f(j|\mathbf{x})$ nonparametrically, perhaps using a nearest neighbour method with the Euclidean metric. This can even be extended to an iterative approach: start with the Euclidean metric to produce a nearest neighbour estimate $\hat{f}(j|\mathbf{x})$, use this estimate to define local regions in terms of almost equal $\hat{f}(j|\mathbf{x})$ and base a new nearest neighbour estimate $\hat{f}(j|\mathbf{x})$ on these regions, and so on. Alternatively, one may note that many classification problems have relatively smooth contours of $f(j|\mathbf{x})$ (hence the success of linear discriminant analysis) and estimate those contours using a global approach (such as linear or quadratic discriminant analysis). Regions along the contours have equal $f(j|\mathbf{x})$ if the approximations are accurate. Distance for use in the nearest neighbour method should therefore be defined as orthogonal to those contours.

To illustrate, suppose that we approximate the contours of $f(j|x)$ using linear discriminant analysis, and suppose that the direction orthogonal to the contours is denoted by $\mathbf{w}$. Then the squared distance between two points, $\mathbf{x}$ and $\mathbf{y}$, is

$$(\mathbf{x} - \mathbf{y})' \mathbf{w} \mathbf{w}' (\mathbf{x} - \mathbf{y})$$

This only measures the component of distance which is orthogonal to the (estimated) equiprobability contours. Of course, $\mathbf{w}$ is only estimated, and the contours are unlikely to be exactly linear in any case. So we relax the rigidity of this approach by including a component of distance along the contour in the metric. That is, we replace the above by

$$(\mathbf{x} - \mathbf{y})' (\mathbf{I} + D\mathbf{w}\mathbf{w}')(\mathbf{x} - \mathbf{y})$$

Here D is a parameter which determines the relative importance of distances parallel to and perpendicular to the estimated contours of $f(j|x)$.

To apply the nearest neighbour method it is necessary to choose a value for k. Also, if the metric choice just outlined is to be used, it is necessary to choose a value for D. They can be estimated from the design data in a manner analogous to the leaving-one-out approach to estimating h in the kernel method; error rate could be used as the criterion. In fact, it often seems to be the case that choice of k is not critical; plots of error rate against k often show a broad flat curve.

Whereas the kernel method uses all the design set points, weighting them according to the kernel function, the nearest neighbour method only uses the k nearest neighbours. This requires a search through the entire design set to identify those k. In general, such searching—and the sorting it inevitably involves—is a time-consuming operation and two broad classes of approaches have been suggested to reduce it.

One approach is based on efficient search strategies such as the branch and bound method. Suppose that we want to find the nearest element of the design set $\{\mathbf{X}_i\}$ to the point $\mathbf{x}$ to be classified. First the design set is decomposed into a hierarchical structure of subsets. To begin with, the entire set may be split into c groups such that members of one group are close to each other, for each group. Then each of these c groups are further decomposed into c subgroups, again so that neighbouring points fall in the same subgroup. And so on. Hierarchical cluster analysis to produce compact spherical clusters is one way of doing this. Finally, the *radii* of each of the groups at the bottom of the tree are determined; for a particular group this is the maximum distance from the mean of that group to any point in that group. Now, consider a group with radius r and mean vector $\mathbf{M}$. Suppose we already know that some design set point $\mathbf{Y}$ is at distance $d(\mathbf{x}, \mathbf{Y})$ from the point $\mathbf{x}$ to be classified. If we find that $d(\mathbf{x}, \mathbf{M}) - r > d(\mathbf{x}, \mathbf{Y})$ then *no point in this subgroup can be closer to $\mathbf{x}$ than $\mathbf{Y}$ is*. This enables whole groups of points to be eliminated in one go, without the need for examining each individual point. The cost of this procedure is that the design set has to be preprocessed by the cluster analysis. But this is rarely a problem as it can be done off-line, before the system is run in a practical application.

The second approach to improving the efficiency of nearest neighbour classification methods also involves preprocessing the design set, though in a radically different way. This class of approaches can also be motivated as a way of improving classification performance by smoothing the decision surface. We shall look at three methods in this class.

The first is the *condensed nearest neighbour approach*, which we illustrate by considering the 1-nearest neighbour method. This method seeks a subset of the design set which classifies the entirety of the design set correctly. To find such a subset, begin by extracting a single design set element and sequentially classifying all the others using the nearest neighbour approach. Eventually one of the other design set points will be misclassified. When this happens, transfer it across to join the single element, classifying the remaining points using the nearest neighbour method relative to this new two-point set. Again, sooner or later, another point is likely to be misclassified. When this occurs, add the misclassified point to the other two. When the entire design set has been processed in this way it will have been partitioned into two: one half will consist of points which were misclassified during the training process and so were transferred, the other half will consist of the remainder. The second half is then processed again, using the first half. This is repeated as often as necessary, until no points are transferred from the second half. The second half then consists of those points which are correctly classified by the remainder. This half is then discarded, leaving a 'condensed' design set.

The motivation for the condensed nearest neighbour method can be seen by considering an area of the measurement space where only points from class j occur. All points falling in this area will be classified as class j because the area has a preponderance of class j design set points. And this would be true even if the design set points in that region were thinned out. Perhaps a tiny fraction of the points in that region would suffice to ensure that all points falling there were classified as class j. If all but this tiny fraction were removed, then less searching through the design set would be required, without any damage to the classification accuracy. The condensed nearest neighbour method is a way of thinning out the design set without damaging its classification accuracy.

The *reduced nearest neighbour approach* is an extension of the condensed nearest neighbour approach. Once the above condensation process is complete, each of the points which have been transferred is examined, one at a time, to see if it is necessary in the design set. If any such point is correctly classified by the other points that have been transferred, it is deemed unnecessary to retain it in the transferred set and it is moved back to the other set. This is unlikely to lead to a great reduction in the numbers of points needed to be retained in the design set, beyond the reduction achieved by the initial condensed nearest neighbour method.

Although the straightforward k-nearest neighbour method can be thought of as providing estimates of $\hat{f}(j|\mathbf{x})$, which are then used to produce a classification by choosing the largest, this approach does not require accurate probability estimates. All that is needed is that the 'estimate' for the jth class leads to the largest value when points should be classified to that class. Put another way, any monotonic increasing function of $\hat{f}(j|\mathbf{x})$ would do as well. The condensed nearest neighbour method capitalises on this by making sure that the j corresponding to the largest $\hat{f}(j|\mathbf{x})$ still corresponds to the largest, even though not all of the design set is used.

We can take this idea further by noting that, in a region where a single (or a few) class j points occur with many class m points, the nearest neighbour estimates will overwhelmingly favour class m. The class j design set points will be swamped and will not affect the classification accuracy. However, they will be retained in the design set by the condensed nearest neighbour process simply because they are lying amid a horde of class m points and will therefore be misclassified by the other design set points.

The *edited nearest neighbour* approach preprocesses the design set to eliminate such sparse class m points lying in regions dense with points from other classes. The effect is to smooth out the sample-based nearest neighbour decision surface, with the aim being to produce a surface which more accurately reflects the underlying true decision surface and is less influenced by sample idiosyncrasies.

Although the above outline was given in terms of $k = 1$, the extension to larger k is straightforward.

These nearest neighbour methods abandon the accuracy of the probability estimates $\hat{f}(j|\mathbf{x})$ while retaining their order at each $\mathbf{x}$ in the design set (or, at least, while retaining the information necessary to classify points in the region to the class corresponding to the largest of the $\hat{f}(j|\mathbf{x})$). Sacrificing this information is not without its disadvantages. Chapter 6 discusses classes of measures of classification performance which are based on the accuracy of the probability estimates, but here we look at another disadvantage. In general the misclassification rate of a rule can be made arbitrarily small by the simple expedient of refusing to classify those objects about which one is not very confident. Such a strategy is sometimes called the *reject option*, not to be confused with *reject inference* (Section 9.5), because one rejects such objects from the classification process.

Now, confidence in one's classification hinges on the size of the probability estimate $\hat{f}(j|\mathbf{x})$. One will be more confident in a classification where $\max_j \hat{f}(j|\mathbf{x})$ is far greater than its nearest competitor than in a classification in which its nearest competitor is almost as large. But if the absolute sizes of the $\hat{f}(j|\mathbf{x})$ estimates have been abandoned, with only their order retained, one cannot determine a degree of confidence. In general, methods which involve preprocessing the design set are at risk from this problem, since they may distort the sampling probabilities of coming from the various classes at each $\mathbf{x}$.

Finally, it is of some interest to note that the g class nearest neighbour asymptotic misclassification rate is bounded above by $e_B\{2 - e_B g/(g-1))\}$, where e_B is the Bayes error rate, and hence is bounded above by twice the Bayes error rate.

5.4   HIGH DIMENSIONAL SPACES

Both kernel and nearest neighbour methods estimate probabilities at a point x using the classes of neighbouring design set points (even if this is via the class conditional distributions and even if the contribution of each neighbouring point is partial, via a kernel function). Bias is likely to be introduced by these methods if the probabilities $f(j|X)$ at neighbouring design set points X differ from the corresponding probability at x. These probabilities are likely to differ less if the X are close to x, hence *nearest* neighbours. However, high dimensions can be deceptive where probability distributions are concerned. For example, with a one-dimensional normal distribution, almost 90% of the probability lies within ± 1.6 standard deviations of the mean. However, with a ten-dimensional spherical multivariate normal distribution, only 1% of the probability lies closer than 1.6 standard deviations from the mean. The mean of such a distribution, though associated with the peak of the density function, is sparsely populated with sample points, most points lying far from the origin (although these regions are sparsely populated too).

To compensate for this effect of the curse of dimensionality, a dramatically larger sample is required to maintain the density of design sample points. Silverman (1986) illustrates this using the example of estimating the density of a multivariate normal distribution at its mean using a normal kernel and the window width which minimises the mean squared error. He shows that to maintain $E[\hat{f}(0) - f(0)]^2/f(0)$ less than 0.1 requires $n = 4$ when $d = 1$, $n = 2790$ when $d = 5$, and $n = 842\,000$ when $d = 10$.

5.5   FURTHER READING

Kernel smoothing methods, especially for probability density functions and regression, have been the subject of intensive research over the past few decades. This is a result of increased computer power, and is demonstrated by the appearance of several recent books on the topic. An early one, Hand (1982), focused on classification as the objective. The most recent, Wand and Jones (1995), is broader in its coverage, including regression and estimation of functionals. This book provides a good overview of recent work using criteria based on mean squared error—reflecting the bulk of work in the area—and describes the h estimation methods outlined above, as well as others. Silverman (1986) focuses on density estimation, and Scott (1992) on multivariate density estimation (though with a chapter on regression). Härdle (1991) describes implementations of smoothing methods in the S statistical language. McLachlan (1992) also includes a good review of work on kernel methods in the context of classification problems.

A large number of papers have been written on kernel methods, too many to review here. Fix and Hodges (1951) was possibly the earliest formal description of kernel/nearest neighbour approaches for classification problems, and M. Rosenblatt (1956) and Parzen (1962) for specifically kernel density estimation; sometimes the phrases *Rosenblatt* or *Parzen kernel method* are used. Epanechnikov (1969) first described the optimal kernel shape which bears his name. Marron and Nolan (1989) describe how to choose canonical kernels which lead to the factorisation of the AMISE. Cacoullos (1966) was perhaps the first to explore product kernels. An important paper describing kernels for categorical data is that of Aitchison and Aitkin (1976).

The condensed nearest neighbour method was proposed by Hart (1968), the reduced nearest neighbour method by Gates (1972) and the edited nearest neighbour method by Hand and Batchelor (1978). Other variants and extensions of these ideas can be developed; see, for example, Chidananda Gowda and Krishna (1979). Fukunaga and Narendra (1975) describe the application of the branch and bound method to nearest neighbour search. The reject option with nearest neighbour methods is developed in considerable detail by Devijver and Kittler (1982). Work on the choice of metric includes that of Short and Fukunaga (1981), Fukunaga and Flick (1984), Todeschini (1989), Myles and Hand (1990) and Henley and Hand (1996). Examples of nearest neighbour methods developed by the machine learning community are given in Aha *et al.* (1991) and Stanfill and Waltz (1986).

For a proof that the asymptotic nearest neighbour error rate is bounded by twice the Bayes error rate see Cover and Hart (1967) or Hand(1981a).

Part III

Evaluating Rules

CHAPTER 6

Aspects of Evaluation

6.1 INTRODUCTION

Superficially, the objective of building a classification rule is straightforward: one wants to classify correctly as many future objects as possible. However, this apparently straightforward aim in fact conceals a number of related objectives and each different objective will correspond to a different performance criterion. This part of the book describes such criteria and how to estimate them.

By far the most popular measure of performance is *error rate* or *misclassification rate*. This is simply the proportion of objects misclassified by the rule. But complications arise even with this most simple of measures. In particular, one would ideally want to avoid testing the rule on the data used for its construction; the rule was, in some sense, optimised for the design data, so that a value of error rate estimated from it should be expected to be optimistically biased as an estimate of future performance. The trouble is that the number of available objects is often limited, and they must be used both for construction and performance evaluation. This has led to the development of subtle methods for error rate estimation, such as the *jackknife* and *bootstrap* methods, described in detail in Chapter 7. Moreover, it is important to be clear about precisely what kind of error rate is being estimated. For example, does one want to know the likely future error rate for a given design set, or is one interested in general comparative performance of a type of classification rule, unconditional on a particular design set?

If error rate is the commonest assessment criterion, this is typically because it is the default criterion: it does not necessarily mean it is the most appropriate criterion. In particular, error rate implicitly assumes that the costs of different types of misclassification are equal (from class 1 to class 2 and vice versa, for example). Needless to say, this is unlikely to be the case in most real applications.

Similarly, error rate says nothing about the relative severity of different misclassifications to the same class: an object misclassified because it is just barely on the wrong side of a threshold will be accorded the same weight as one misclassified because it is well away from the threshold, on the wrong side. In some situations this is the appropriate thing to do, but in others a measure of relative severity would provide useful information.

Error rate also tells us nothing about the accuracy of the probability estimates implicit in the classifier. If the threshold is not determined a priori for all future applications (perhaps because priors or costs vary with time) then one will want these probability estimates to be accurate over a range of potential thresholds, not simply at one point.

Clearly the distinctions between these different aspects of classifier performance are subtle and careful definition is needed. The next section outlines our terminology; this differs from the terminology that others have used, and the reasons for introducing new terms are given in Section 6.10.

Classification problems often involve constraints which change their nature. An example arises from the distinction between *diagnosis* and *screening* in medicine (see Section 8.2). Diagnosis requires one to classify a subject into the class from which he or she has the highest estimated probability of coming (subject to cost considerations). But screening is a *population* problem: the aim is to identify that fraction of the population most likely to be suffering from the disease. Given that one knows (or has an estimate of) the size of this fraction beforehand, this problem will not be solved by separate classification of individuals. The consequence is that, although error rate may still be used to assess performance, it is error rate from a restricted class of classifiers. A similar situation arises in automated chromosome analysis, as described in Section 10.2.

Finally, error rate may simply be inappropriate. In *prevalence estimation*, a classification rule is used to assign subjects to different strata, a sample from each of which is then accurately classified. The two classifications are then combined to yield an estimate of prevalence. And the aim here is to obtain as *accurate* as possible an estimate of prevalence, measured by a criterion such as variance not error rate. Again, this is discussed in Section 8.2.

Although error rate is overwhelmingly the most popular criterion for comparison of classification rules, all too often it does not answer the researcher's real question. The use of error rate often suggests insufficiently careful thought about the real objectives of the research, a situation described in Hand (1994a).

Classifier performance may be assessed for two basic reasons: to *compare* classifiers (is this one better than that one?) or to determine an *absolute measure of quality* of performance (is the classifier good enough for the problem at hand?). When comparing classifiers it is necessary to take account of the fact that common data sets may be used by the two classifiers; the 'two sets' of objects being classified are not independent.

To keep things simple initially, in the next section we assume that we have a test set $x_1, ..., x_n$, independent of the design set.

First we clarify some notation. As in preceding chapters, we will use f as the generic form for a probability distribution. Hence $f(x, j)$ represents the joint probability of measurement vector x and class j, $f(j|x)$ represents the conditional probability that an object with measurement vector x belongs to class j, and so on. We will let $\hat{f}$ represent the estimated distribution given by the classification rule in question.

6.2 SOME DEFINITIONS

We suppose that we have a classification rule, based on d measurements, and that we have available a test set consisting of n objects, on each of which the d measurements have also been taken. Let c_i and x_i be, respectively, the true class (taking values 0, 1, 2, ..., C) and the measurement vector for the ith test object. Let x be an arbitrary measurement vector. Then we can define four concepts:

- *Inaccuracy*: how (in)effective the rule is in assigning an object to the correct class. It will be based on a measure of difference between the true class and the estimated probability of belonging to that class. For example, based on a summary of the $|\delta(j|x_i) - \hat{f}(j|x_i)|$, where $\hat{f}(j|x_i)$ is the rule's estimated probability that the ith object belongs to class j and $\delta(j|x_i)$ is 1 if $c_i = j$ and 0 otherwise.
- *Imprecision*: how different the estimated probabilities $\hat{f}(j|x)$ are from the true probabilities $f(j|x)$. In principle, such a measure could be based on the differences $f(j|x) - \hat{f}(j|x)$, although since $f(j|x)$ is unknown, some subtleties will be required.
- *Inseparability*: how similar are the true probabilities of belonging to each class at x, averaged over x. If the probabilities at x are similar, the distribution across classes at x is not dominated by any one class—the situation is inseparable. Since we are hoping for clear differences between the classes, inseparability is to be avoided. We are aiming for separability; ideally all objects with a given x will belong to a single class. An example of a possible element of an inseparability index, measured at x, is the negative of the variance of the $f(j|x)$:

$$-V_j(f(j|x)) = -\sum_j \left\{ f(j|x) - \sum_j f(j|x)^2/C \right\}^2$$

which will take low values when the probabilities $f(j|x)$ are dissimilar. This will be averaged over x to yield an overall measure. The negative sign implies that *low* values are *good*, an interpretation which we prefer, for reasons given below.

- *Resemblance*: resemblance measures the variation between the true probabilities conditioned on the estimated ones. We would like these probabilities to be very different. Letting $\hat{\mathbf{f}}(\mathbf{j}|x) = (\hat{f}(1|x), ..., \hat{f}(C|x))$, we want to know how different are the elements of the vector $\mathbf{f}(\mathbf{j}|\hat{\mathbf{f}}(\mathbf{j}|x)) = (\mathbf{f}(1|\hat{\mathbf{f}}(\mathbf{j}|x)), ..., \mathbf{f}(C|\hat{\mathbf{f}}(\mathbf{j}|x)))$. An example of a possible measure would be the the the variance of the $f(k|\hat{\mathbf{f}}(\mathbf{j}|x))$, telling us how different they are from each other. For the same reason as with inseparability, we would use the negative of the variance—low values (corresponding to large variance) are good.

To summarise, we want high accuracy, high precision and high separability, hence low values of the inaccuracy coefficients, low values of the imprecision coefficients, and low values of the inseparability coefficients. We also want low resemblance, hence low values of the resemblance coefficients. This is the reason for the negative sign in our definitions of inseparability and resemblance coefficients. It means that, for *all* of the measures, *low* values correspond to *good*. We shall see below that important special cases of imprecision and inaccuracy are related, and they are both positive, with low values being desirable. Since low values are desirable, imprecision and inaccuracy seemed semantically more attractive terms than the superficially simpler precision and accuracy. For consistency, so that low is desirable in all our measures, we adopted inseparability and defined both inseparability and resemblance so that low is also good. Section 6.10 has more to say about terminology.

These concepts answer some of the questions raised in Section 6.1. For example, error rate is a type of inaccuracy measure. This can be seen for the two-class case ($j = 0, 1; c_i = 0, 1$) by summarising the $|\delta(j|\mathbf{x}_i) - \hat{f}(j|\mathbf{x}_i)|$ using

$$\frac{1}{n}\sum_i \mathrm{I}\left\{|c_i - \hat{f}(1|\mathbf{x}_i)| - 0.5\right\}$$

which is error rate (where $\mathrm{I}(x)$ is 1 for positive x and 0 otherwise).

Note that inseparability, as defined above, is not a property of the classification rule (except insofar as the true probabilities are determined by the choice of variables). This means that inseparability is a rather different concept from impurity, as used in Chapter 4. We will discuss this difference below, where we will also include resemblance in the comparison.

An effective classifier will have low imprecision and resemblance coefficients. Both may be needed. Low imprecision alone will not mean low error rates if the true probabilities throughout the space are near to the priors and these priors are almost equal. Low resemblance means that there are real differences between the true probabilities, and the classification rule could possibly take advantage of them; but it may not be able to if the rule is imprecise.

6.3   INACCURACY

In Section 6.2 we showed that error rate was a kind of inaccuracy measure. But in Section 6.1 we noted how error rate had the (possible) disadvantage that it did not take any account of how far from the thresholds the estimated probabilities of class memberships lay. Other measures of inaccuracy have been defined which overcome this weakness. One common measure (and one which is probably second in terms of popularity to error rate and which is often used as a criterion for neural network optimisation) is the *Brier* or *quadratic* score. This is defined as

$$\frac{1}{n}\sum_{i=1}^{n}\sum_{j}\left\{\delta(j|\mathbf{x}_i)-\hat{f}(j|\mathbf{x}_i)\right\}^2$$

which may equivalently be written

$$\frac{1}{n}\sum_{i=1}^{n}\left\{(1-\hat{f}(c_i|\mathbf{x}_i))^2+\sum_{j\neq c_i}\hat{f}(j|\mathbf{x}_i)^2\right\}$$

and

$$\frac{1}{n}\sum_{i=1}^{n}\sum_{j}\left[\delta(j|\mathbf{x}_i)\left\{1-\hat{f}(j|\mathbf{x}_i)\right\}^2+(1-\delta(j|\mathbf{x}_i))\hat{f}(j|\mathbf{x}_i)^2\right]$$

For the two-class case this is equivalent to

$$\frac{2}{n}\sum_{i=1}^{n}\left\{c_i-\hat{f}(1|\mathbf{x}_i)\right\}^2$$

Being a sum of squared deviations, it has a natural interpretation in terms of mean squared error. We shall have more to say about the Brier score below, but here we shall merely note that, when $\hat{f}(j|\mathbf{x})$ takes only the values 0 and 1 (i.e. the probability that an object at $\mathbf{x}$ belongs to some class is estimated to be 1 and the probability that it belongs to the other classes is estimated to be 0), the Brier score is equivalent to error rate.

Another measure which has obvious statistical appeal is the logarithmic score

$$-\frac{1}{n}\sum_{i=1}^{n}\sum_{j}\delta(j|\mathbf{x}_i)\ln\hat{f}(j|\mathbf{x}_i)=-\frac{1}{n}\sum_{i=1}^{n}\ln\hat{f}(c_i|\mathbf{x}_i)$$

which is seen to be (minus) the log-likelihood. In the two-class case it can be expressed in the form

$$-\frac{1}{n}\sum_{i=1}^{n}\left\{c_i\ln\hat{f}(1|\mathbf{x}_i)+(1-c_i)\ln\left(1-\hat{f}(1|\mathbf{x}_i)\right)\right\}$$

Measures of inaccuracy are sometimes called *scoring rules*, since they yield overall scores of the quality of the $\hat{f}(j|\mathbf{x})$ estimates. They are also sometimes called *quasi-utility* functions or *pseudo-utility* functions since they give a mathematical formalisation of a utility function for the classification rule.

The most general definition of an inaccuracy measure is that it is a measure of distance between the empirical test set function $\delta(j|\mathbf{x}_i)$ and the classifier function $\hat{f}(c_i|\mathbf{x}_i)$. A general class of such measures can be written in the form

$$In=\sum_{i}\sum_{j}g\left[h(\delta(j|\mathbf{x}_i)),h\left(\hat{f}(j|\mathbf{x}_i)\right)\right]$$

g here can be regarded as a loss function; the aim is to find that $\hat{f}$ which minimises it. h is some transformation, such as logarithm in the above.

Since one can easily devise inaccuracy measures, it is natural to seek criteria by which one might choose between them. One such criterion is based on the idea of *strictly proper* rules. The expected contribution to the general inaccuracy measure given by the ith object is

$$E\left(\sum_j g\left[h(\delta(j|\mathbf{x}_i)), h\left(\hat{f}(j|\mathbf{x}_i)\right)\right]\right) = \sum_j f(j|\mathbf{x}_i)g\left[h\left(\delta(j|\mathbf{x}_i)\right), h\left(\hat{f}(j|\mathbf{x}_i)\right)\right]$$

A strictly proper measure is then one which satisfies

$$\sum_j f(j|\mathbf{x}_i)g[h\,(\delta(j|\mathbf{x}_i)), h(f(j|\mathbf{x}_i))] < \sum_j f(j|\mathbf{x}_i)g\left[h(\delta(j|\mathbf{x}_i)), h(\hat{f}(j|\mathbf{x}_i))\right]$$

That is, the minimum loss of a strictly proper measure is attained when the true function is used; if assessed by a proper measure, no classifier can do better than one based on the true probabilities. (A nonstrictly proper measure has $<$ replaced by $\leq$.)

To see why this concept is useful, consider the proposed inaccuracy measure

$$In = -\frac{1}{n}\sum_i \sum_j \hat{f}(j|\mathbf{x}_i)\delta(j|\mathbf{x}_i)$$

based on

$$g\left[h(\delta(j|\mathbf{x}_i)), h\left(\hat{f}(j|\mathbf{x}_i)\right)\right] = -\hat{f}(j|\mathbf{x}_i)\delta(j|\mathbf{x}_i)$$

Superficially this looks good: the larger the value of $\hat{f}$ assigned to the true class, the better (it leads to lower values of In). However, this measure is not proper; it allows us to produce a classifier which appears to perform *better* than the classifier based on the true probabilities. Any system which has $\hat{f}(j|\mathbf{x}_i)\delta(j|\mathbf{x}_i) > f(j|\mathbf{x}_i)\delta(j|\mathbf{x}_i)$ will appear to do better. That is, any system which assigns more than the true probability to the true class will appear to do better. In a word, the system will be *overconfident*. Proper measures thus favour precise classifiers. Both Brier score and logarithmic score are strictly proper. Error rate is nonstrictly proper.

6.4   IMPRECISION

We see from the definition at the beginning of Section 6.2 that a rule will be deemed precise (low imprecision) at $\mathbf{x}$ if the estimated probability of belonging to class j at $\mathbf{x}$ is approximately equal to the proportion of objects at $\mathbf{x}$ which belong to class j. Whether imprecision matters or not depends on the application. If, in a two-class problem, it has been predetermined that a particular proportion will be classified as class 0 then all that matters is that the estimated probabilities of belonging to class 0 should be roughly monotonically related to the true probabilities. In this situation precision is unimportant (an example is given in Section 8.2). But accuracy is important if, more generally, the estimated probability is to be compared with a predetermined threshold (for example, if all objects with estimated probability greater than 1/2 of belonging to class 0 are to be classified as class 0). Moreover, if the threshold is not predetermined— perhaps because the prior probabilities and/or costs are not fixed—the accuracy matters throughout a range of possible values of the estimated probability. Also, our confidence in a classification system will be reinforced if we know that, of those objects to which we assign a probability of 0.8 of belonging to class 0, 80% really do belong to class 0. This can be vital in systems which defer decisions on doubtful objects while more information is collected (more variables measured).

Just as inaccuracy measures compare $\delta(j|\mathbf{x}_i)$ with $\hat{f}(j|\mathbf{x}_i)$, so imprecision measures compare $f(j|\mathbf{x}_i)$ with $\hat{f}(j|\mathbf{x}_i)$. The problem is that we do not know the true probabilities. (Otherwise we would have no need to estimate them!) Moreover, at any given $\mathbf{x}$ there are likely to be very few sample points from which to obtain an accurate estimate of $f(j|\mathbf{x})$ (since $\mathbf{x}$ will often be continuous and multivariate). However, we do have a sample from $f(\mathbf{x},j)$, namely the test set sample. Using this, the basic strategy is to compute some *summary statistics* from $\hat{f}(j|\mathbf{x})$ and to compare them with the corresponding estimated summary statistics computed from the test sample.

Two questions then arise: What summary statistics should be used? And how should they be compared? Different summary statistics probe different aspects of the difference between $\hat{f}(j|\mathbf{x})$ and $f(j|\mathbf{x})$. The problem has been likened to testing a pseudorandom number generator: there is an infinite number of tests which could be carried out. Successfully passing a test reinforces one's belief that the classifier is precise, but this does not make it a fact—the classifier may fail some other test.

In an attempt to overcome this problem, comparison of separate test statistics in this way has been extended by combining the results of several such comparisons into a single test statistic. For example, one might adopt statistical moments as the summary statistics and combine them into a single overall measure. Alternatively, one might compute low-order multivariate Fourier coefficients and compute the sum of squared differences between the coefficients obtained from the $\delta(j|\mathbf{x}_i)$ and the $\hat{f}(j|\mathbf{x}_i)$. In effect, one is computing an estimate of the $f(j|\mathbf{x})$ from the test set and comparing this estimate with the estimate obtained from the design set. If one adopts a *complete* set of basis functions with which to estimate $f(j|\mathbf{x})$ from the test set, the estimate used in the comparisons will be estimated by the empirical test set function $\delta(j|\mathbf{x}_i)$ itself. This means that one will be comparing $\hat{f}(j|\mathbf{x})$ with the empirical sample distribution; one will, in fact, be producing a measure of *inaccuracy*, not imprecision. By taking only a limited subset of the functions (be they statistical moments, Fourier coefficients or whatever) one is introducing a degree of smoothing into the estimate of $f(j|\mathbf{x})$ and this smoothing allows us to estimate imprecision.

One way of obtaining suitable summary statistics is to compute an *integral transformation* of the $f(\mathbf{x}, j)$ function:

$$\int \phi_{\alpha j}(\mathbf{x}) f(\mathbf{x}, j) d\mathbf{x} = \int \phi_{\alpha j}(\mathbf{x}) f(j|\mathbf{x}) f(\mathbf{x}) \, d\mathbf{x}$$

Choice of ϕ will determine what transformation, what statistic, we compute.

Through what is sometimes called the *plug-in principle*, in which an estimator is obtained by replacing f with the empirical distribution δ, the empirical sample statistic is

$$S_{\alpha j} = \frac{1}{n} \sum_{i=1}^{n} \{ \phi_{\alpha j}(\mathbf{x}_i) \delta(j|\mathbf{x}_i) \} \tag{6.1}$$

and the estimate of the same statistic computed from the classifier function is

$$\hat{S}_{\alpha j} = \int \phi_{\alpha j}(\mathbf{x}) \hat{f}(j|\mathbf{x}) f(\mathbf{x}) d\mathbf{x}$$

$$= \frac{1}{n} \sum_{i=1}^{n} \{ \phi_{\alpha j}(\mathbf{x}_i) \hat{f}(j|\mathbf{x}_i) \} \tag{6.2}$$

An overall imprecision measure will then be obtained by comparing them and combining the comparisons over the classes. An obvious way to compare them is simply to take their difference. Taking the difference of (6.1) and (6.2) and summing over classes yields

$$R_\alpha = \sum_j (S_{\alpha j} - \hat{S}_{\alpha j}) = \sum_j \frac{1}{n} \sum_{i=1}^{n} \left\{ \phi_{\alpha j}(\mathbf{x}_i) \left[\delta(j|\mathbf{x}_i) - \hat{f}(j|\mathbf{x}_i) \right] \right\}$$

It can be viewed as a transformation of the difference between the empirical and classifier functions, which is just what we want from an estimate of imprecision. Put another way, this method compares $E\left[\phi(\mathbf{x}) \hat{f}(\mathbf{x}) \right]$ with $E[\phi(\mathbf{x}) \delta(\mathbf{x})]$.

We can also generalise this approach by replacing $\delta(j|\mathbf{x}_i)$ and $\hat{f}(j|\mathbf{x}_i)$ in (6.1) and (6.2) by some functions, $H_j(\delta(j|\mathbf{x}_i))$ and $H_j(\hat{f}(j|\mathbf{x}_i))$, of them. We give an example of this in the next section.

Example 6.1 Consider the case of two classes, $j = 0, 1$ and let $\phi_{\alpha j}(\mathbf{x}_i) = (1 - \hat{f}(j|\mathbf{x}_i))^2$.

Then

$$S_\alpha = \sum_j S_{\alpha j} = \frac{1}{n} \sum_{i=1}^{n} \left\{ c_i \hat{f}(0|\mathbf{x}_i)^2 + (1 - c_i) \hat{f}(1|\mathbf{x}_i)^2 \right\}$$

and

$$\hat{S}_\alpha = \sum_i \hat{S}_{\alpha j} = \frac{1}{n}\sum_{i=1}^{n}\left\{\hat{f}(1|\mathbf{x}_i)\hat{f}(0|\mathbf{x}_i)^2 + \hat{f}(0|\mathbf{x}_i)\,\hat{f}(1|\mathbf{x}_i)^2\right\}$$

In general, for more than two classes, we have

$$S_\alpha = \sum_j S_{\alpha j} = \frac{1}{n}\sum_{i=1}^{n}\sum_j \delta(j|\mathbf{x}_i)\{1 - \hat{f}(j|\mathbf{x}_i)\}^2$$

and

$$\hat{S}_\alpha = \sum_j \hat{S}_{\alpha j} = \frac{1}{n}\sum_{i=1}^{n}\sum_j \hat{f}(j|\mathbf{x}_i)\left\{1 - \hat{f}(j|\mathbf{x}_i)\right\}^2$$

Example 6.2 Again with two classes, let $\phi_{\alpha 0}(\mathbf{x}_i) = -\ln \hat{f}(0|\mathbf{x}_i)$ and $\phi_{\alpha 1}(\mathbf{x}_i) = -\ln \hat{f}(1|\mathbf{x}_i)$.

Then

$$S_\alpha = \sum_j S_{\alpha j} = -\frac{1}{n}\sum_{i=1}^{n}\left\{c_i \ln \hat{f}(1|\mathbf{x}_i) + (1 - c_i) \ln \hat{f}(0|\mathbf{x}_i)\right\}$$

and

$$\hat{S}_\alpha = \sum_j \hat{S}_{\alpha j} = -\frac{1}{n}\sum_{i=1}^{n}\left\{\hat{f}(1|\mathbf{x}_i) \ln \hat{f}(1|x_i) + \hat{f}(0|\mathbf{x}_i) \ln f(0|\mathbf{x}_i)\right\}$$

The difference between these is

$$R_\alpha = \frac{1}{n}\sum_{i=1}^{n}\left(c_i - \hat{f}(1|\mathbf{x}_i)\right)\ln\left(\hat{f}(1|\mathbf{x}_i)/\hat{f}(0|\mathbf{x}_i)\right)$$

In general, for more than two classes, these are

$$S_\alpha = \sum_j S_{\alpha j} = -\frac{1}{n}\sum_{i=1}^{n}\sum_j \delta(j|\mathbf{x}_i) \ln \hat{f}(j|\mathbf{x}_i)$$

and

$$\hat{S}_\alpha = \sum_j \hat{S}_{\alpha j} = -\frac{1}{n}\sum_{i=1}^{n}\sum_j \hat{f}(j|\mathbf{x}_i) \ln \hat{f}(j|\mathbf{x}_i)$$

Example 6.3 Note that the S_α in Example 6.1 is not quite the same as the Brier score since only the terms corresponding to the true class contribute to the sum. (However, in the special case of two classes it reduces to half of the Brier score.) We showed above that the Brier score could be expressed as

$$\frac{1}{n}\sum_{i=1}^{n}\sum_j \left[\delta(j|\mathbf{x}_i)\left\{1 - \hat{f}(j|\mathbf{x}_i)^2\right\} + (1 - \delta(j|\mathbf{x}_i))\,\hat{f}(j|\mathbf{x}_i)^2\right]$$

This can alternatively be written as

$$T_\alpha = \frac{1}{n} \sum_{i=1}^{n} \sum_{j} \left[\delta(j|\mathbf{x}_i)\left\{ 1 - 2\hat{f}(j|\mathbf{x}_i) \right\} + \hat{f}(j|\mathbf{x}_i)^2 \right]$$

(which cannot be expressed as an integral transformation, as above, because of the second term on the right-hand side). Replacing $\delta(j|\mathbf{x}_i)$ in this by $\hat{f}(j|\mathbf{x})$ leads to

$$\hat{T}_\alpha = \frac{1}{n} \sum_{i=1}^{n} \sum_{j} \left[\hat{f}(j|\mathbf{x}_i)\left\{ 1 - 2\hat{f}(j|\mathbf{x}_i) \right\} + \hat{f}(j|\mathbf{x}_i)^2 \right]$$

so that

$$T_\alpha - \hat{T}_\alpha = \frac{1}{n} \sum_{i=1}^{n} \sum_{j} \left[\delta(j|\mathbf{x}_i) - \hat{f}(j|\mathbf{x}_i) \right]\left\{ 1 - 2\hat{f}(j|\mathbf{x}_i) \right\}$$

We see that this measure of imprecision, derived from the Brier score, is in fact based on integral transformations S_α and $\hat{S}_\alpha$ using $\phi_{\alpha j}(\mathbf{x}_i) = \left\{ 1 - 2\hat{f}(j|\mathbf{x}_i) \right\}$

One way of interpreting imprecision is to regard $S_{\alpha j}$ as an observed statistic and $\hat{S}_{\alpha j}$ as its hypothesised value under the hypothesis that $\hat{f}(j|\mathbf{x})$ is the true distribution. To *test* this difference, to see if there is a significant difference between the two estimates (equivalently, to see if the classifier distribution may be regarded as precise), we need to compare the difference with its standard deviation. For the Brier score with two classes, Example 6.3, this is obtained from

$$\text{var}\left\{ \left(c - \hat{f}(1|\mathbf{x}) \right)\left(1 - 2\hat{f}(1|\mathbf{x}) \right) \right\} = \text{var}\left\{ c\left(1 - 2\hat{f}(1|\mathbf{x}) \right) \right\}$$
$$= \left(1 - 2\hat{f}(1|\mathbf{x}) \right)^2 f(1|\mathbf{x})f(0|\mathbf{x})$$

where we have dropped the subscripts for convenience. Finally, estimating $f(j|\mathbf{x})$ by $\hat{f}(j|\mathbf{x})$, a test for the measure of imprecision defined from the Brier score is given by comparing

$$\frac{\sum_{i=1}^{n} \left(c_i - \hat{f}(1|\mathbf{x}) \right)\left(1 - 2\hat{f}(1|\mathbf{x}) \right)}{\sqrt{\sum_{i=1}^{n} \left(1 - 2\hat{f}(1|\mathbf{x}) \right)^2 \hat{f}(1|\mathbf{x})\hat{f}(0|\mathbf{x})}}$$

with a normal distribution. Test statistics for the measure of imprecision in Examples 6.1 and 6.2 can be defined similarly; consider Example 6.2 with two classes:

$$\text{var}\left\{ \left(c - \hat{f}(1|\mathbf{x}) \right) \ln\left(\hat{f}(1|\mathbf{x})/\hat{f}(0|\mathbf{x}) \right) \right\} = f(1|\mathbf{x})f(0|\mathbf{x})\left[\ln\left(\hat{f}(1|\mathbf{x})/\hat{f}(0|\mathbf{x}) \right) \right]^2$$

6.5 AN IMPORTANT IMPRECISION MEASURE

We have

$$
E_x \sum_j E_{\delta|j,x}\left(\delta(j|\mathbf{x}) - \hat{f}(j|\mathbf{x})\right)^2
$$

$$
= E_x \sum_j E_{\delta|j,x}\left[\delta(j|\mathbf{x}) - f(j|\mathbf{x}) - \hat{f}(j|\mathbf{x})\right]^2
$$

$$
= E_x \sum_j E_{\delta|j,x}(\delta(j|\mathbf{x}) - f(j|\mathbf{x}))^2 + E_x \sum_j E_{\delta|j,x}\left(f(j|\mathbf{x}) - \hat{f}(j|\mathbf{x})\right)^2
$$

$$
= E_x \sum_j (1 - f(j|\mathbf{x}))^2 f(j|\mathbf{x}) + E_x \sum_j (0 - f(j|\mathbf{x}))^2 (1 - f(j|\mathbf{x})) \qquad (6.3)
$$

$$
+ E_x \sum_j E_{\delta|j,x}\left(f(j|\mathbf{x}) - \hat{f}(j|\mathbf{x})\right)^2
$$

$$
= E_x \sum_j (1 - f(j|\mathbf{x})) f(j|\mathbf{x}) + E_x \sum_j \left(f(j|\mathbf{x}) - \hat{f}(j|\mathbf{x})\right)^2
$$

The term on the left measures the difference between $\delta(j|\mathbf{x})$ and $\hat{f}(j|\mathbf{x})$ and is the Brier *inaccuracy*. The second term in the final line measures the difference between $\hat{f}(j|\mathbf{x})$ and $f(j|\mathbf{x})$; it is the *imprecision*. When we have two classes, the first term in the final line can be rewritten as

$$
E_x \sum_j (1 - f(1|\mathbf{x})) f(1|\mathbf{x}) = E_x 2 f(0|\mathbf{x}) f(1|\mathbf{x})
$$

$$
= -E_x \left\{ \sum_j (f(j|\mathbf{x}) - \bar{f})^{-2} - \frac{1}{2} \right\}
$$

where $\bar{f} = \{f(0|\mathbf{x}) + f(1|\mathbf{x})\}/2 = \frac{1}{2}$ is the mean of the $f(j|\mathbf{x})$. That is, the first term in the final line of (6.3) is equivalent (for our purposes) to the negative of the variance of the $f(j|\mathbf{x})$. It is thus a measure of inseparability—low values are good.

Summarising, we have the following decomposition (but only for the two-class case):

Inaccuracy = Inseparability + Imprecision

For two or more classes, the Brier inaccuracy score can be estimated directly from the data as $\frac{1}{n}\sum_i \sum_j (\delta(j|\mathbf{x}_i) - \hat{f}(j|\mathbf{x}_i))^2$. Imprecision, involving the unknown true f, cannot be computed from the data. Also, the $\sum_j (1 - f(j|\mathbf{x})) f(j|\mathbf{x})$ term (corresponding to inseparability in the two-class case) is unknown. This means that this decomposition cannot be used to determine the absolute value of the imprecision. However, the $\sum_j (1 - f(j|\mathbf{x})) f(j|\mathbf{x})$ term, although unknown, is a function *solely* of the unknown true $f(j|\mathbf{x})$. This means that it will be the same for all classifiers evaluated on the same test set. Consequently, it will cancel if inaccuracy measures for two classifiers are subtracted. That is, the imprecision of two rules can be *compared* by subtracting their probability scores.

This leaves open the question of how to estimate absolute measures of imprecision. An important class of approaches arises by grouping into cells the space spanned by $\mathbf{x}$. Then the average values of $\hat{f}(j|\mathbf{x})$ in each cell can be compared with the average proportion of test set objects in that cell which fall into class j. For example, this could be done by partitioning the measurement space in terms of the values of $\mathbf{x}$. Or it could be done by conditioning on the $\hat{f}(j|\mathbf{x})$ and partitioning the result. An important special case of this arises with tree classifiers (Chapter 4) where the basic construction of the classification rule produces a natural partition into cells in the space via the $\hat{f}(j|\mathbf{x})$.

Partitioning the measurement space in terms of the values of $\hat{f}(j|\mathbf{x})$ is widely used in assessment of probability forecasts in Bayesian statistics. Bayesian probability forecasts are a very important special case of this discussion because they typically involve the probability that some event occurs, so they are fundamentally two-class problems. Partitioning measures introduce the requisite smoothing into the estimate obtained from the empirical distribution in two ways. Firstly, they group similar $\hat{f}(j|\mathbf{x})$ together into cells, and secondly they average over all the $f(j|\mathbf{x})$ within each cell. The departures from precision that these functions will fail to detect arise in situations when $f(j|\mathbf{x}) \neq \hat{f}(j|\mathbf{x})$ but $f\left(j|\hat{f}(j|\mathbf{x})\right) = \hat{f}(j|\hat{f}(j|\mathbf{x}))$. This means that a method which is precise using this approach will enable one to make statements such as, 'Of the objects which the rule says have an 80% chance of being in class 1, 80% really are.' However, it does not allow one to make statements such as, 'If the rule asserts that 80% of the objects with measurement vector $\mathbf{x}$ are in class 1 then 80% are.'

With this grouping of the $\hat{f}(j|\mathbf{x})$ into cells, let $\hat{p}_{\alpha j}$ denote the proportion in cell α that the classifier estimates belong to class j, let $p^*_{\alpha j}$ denote the proportion of test set elements in cell α that really do belong to class j, let $p_{\alpha j}$ denote the true probability that an object in cell α belongs to class j, and let v_α denote the number of objects which fall in cell α. Then we have the (grouped) Brier inaccuracy

$$E \sum_\alpha \sum_j v_\alpha \left[p^*_{\alpha j}\left(1 - \hat{p}_{\alpha j}\right)^2 + (1 - p^*_{\alpha j})\hat{p}^2_{\alpha j} \right]$$

$$= E \sum_\alpha \sum_j v_\alpha \left[p^*_{\alpha j}\left(1 - p_{\alpha j}\right)^2 + (1 - p^*_{\alpha j})p^2_{\alpha j} \right] E \sum_\alpha \sum_j v_\alpha (p_{\alpha j} - \hat{p}_{\alpha j})^2$$

$$= E \sum_\alpha \sum_j v_\alpha \left(p^*_{\alpha j} - p_{\alpha j} \right)^2 + E \sum_\alpha \sum_j v_\alpha p^*_{\alpha j}(1 - p^*_{\alpha j}) + E \sum_\alpha \sum_j v_\alpha \left(p_{\alpha j} - \hat{p}_{\alpha j} \right)^2$$

$$= E \sum_\alpha \sum_j v_\alpha \left(p^*_{\alpha j} - \hat{p}_{\alpha j} \right)^2 + E \sum_\alpha \sum_j v_\alpha p^*_{\alpha j}(1 - p^*_{\alpha j}) \qquad (6.4)$$

With this grouping strategy, we can now estimate the absolute imprecision directly from the term $E \sum_\alpha \sum_j v_\alpha (p^*_{\alpha j} - \hat{p}_{\alpha j})^2$.

Grouping of objects in this way is useful because, apart from combining the $p_{\alpha j}^* - \hat{p}_{\alpha j}$ to yield overall measures of the quality of a classification rule, we can also look at these differences individually, to find where and in what way the classifier is poor. More generally, we can define other local measures of discrepancy between the classifier and the empirical distribution obtained from the test set.

Since the $p_{\alpha j}^*$ are based on v_α points, which may well differ from cell to cell, comparisons between the $p_{\alpha j}^* - \hat{p}_{\alpha j}$ should make allowance for this. We can therefore standardise the differences to give $X_{\alpha j} = (p_{\alpha j}^* - \hat{p}_{\alpha j})/$ $\sqrt{\hat{p}_{\alpha j}(1 - \hat{p}_{\alpha j})/v_\alpha}$, the *Pearson residuals*. (The sum of squares of these residuals is the usual Pearson χ^2 goodness of fit statistic, summarising the overall quality of the classifier.)

In Example 6.2 we transformed by taking logarithms and we can do the same for the summarising $p_{\alpha j}^*$ and $\hat{p}_{\alpha j}$. In fact, the *deviance residual* is defined as

$$sgn(p_{\alpha j}^* - \hat{p}_{\alpha j})\sqrt{v_\alpha}\left[2p_{\alpha j}^* \ln (p_{\alpha j}^*/\hat{p}_{\alpha j}) + 2(1 - p_{\alpha j}^*) \ln\left((1 - p_{\alpha j}^*)/(1 - \hat{p}_{\alpha j})\right)\right]^{1/2}$$

The sum of squares of these residuals gives the overall deviance of the model:

$$D = 2\sum_\alpha \sum_j v_\alpha \left\{ p_{\alpha j}^* \ln (p_{\alpha j}^*/\hat{p}_{\alpha j}) + (1 - p_{\alpha j}^*) \ln \left((1 - p_{\alpha j}^*)/(1 - \hat{p}_{\alpha j})\right)\right\}$$

6.6 INSEPARABILITY AND RESEMBLANCE

Separability tells us how different are the true probabilities of belonging to each class. Separability should be high (inseparability low) when the probabilities for every measurement vector $\mathbf{x}$ are all 0 except for that corresponding to one class, which (obviously) has the value 1, since this means that, at $\mathbf{x}$, the object is certain to belong to a particular class. That is, separability should be high when the classes are perfectly separated in the measurement space—when the distributions have little or no overlap. A possible measure of inseparability is the negative of the variance of the $f(j|\mathbf{x})$, which is equivalent in the two-class case to $E_x \sum_j (1 - f(j|\mathbf{x}))f(j|\mathbf{x})$. Note again that a *low* score on the inseparability measure is a *good* thing. If we have a set of variables for which most of the $f(j|\mathbf{x})$ are near 0 or 1, we can in principle build a highly effective classifier. A set of variables for which the $f(j|\mathbf{x})$ are seldom near 0 or 1 will mean that, however clever we are in building a classifier on those variables, we will not be able to achieve a highly accurate result (although we may have good precision).

Inseparability is all very well, but it is of limited value in a practical context. In practice, we want to know how different are the *predicted* classes are, Does the predicted classification induced by the rule *really* serve to separate the true classes well? We call this property *resemblance*.

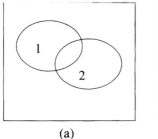
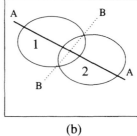

                    (a)   (b)

Figure 6.1: The difference between inseparability and resemblance. See text for explanation.

The distinction and the relationship between inseparability and resemblance are illustrated in Figure 6.1. Suppose that the ellipses show probability contours for two equiprobable classes, with $f(\mathbf{x}|j)$ being constant within each ellipse and zero outside it. The two classes in Figure 6.1(a) are well separated; the situation has a low inseparability value. Now consider the decision surface AA in Figure 6.1(b), which is the decision surface from a poor classifier. Suppose that $\hat{f}(j|\mathbf{x})$ is constant in the region above AA and constant in the region below AA (but with different values in the two regions; we know this from the fact that the decision surface is located where it is). Suppose also that the estimated values of $\hat{f}(j|\mathbf{x})$ in the two regions are very different. Then the rule is confident that it is separating the classes well (because the probability estimates are very different). However, the resemblance of the resulting classification rule is poor (it takes high values) because the true probability of belonging to class 1 in the region above the decision surface in fact does not differ substantially from the true probability of belonging to class 2 in that region. So, inseparability is good (low) but resemblance is poor (high).

Now suppose that the priors are not equal; in particular, suppose that the ellipse for class 1 is much more densely populated than the ellipse for class 2. Now the resemblance will be low (good) since it does separate the classes well, merely because above the line AA, more objects belong to class 1 than class 2. (And below the line AA more objects belong to class 1 than class 2.) Most classifiers would have good resemblance. However, even better resemblance (an even lower value) than that produced by AA would be obtained by the rule corresponding to the decision surface BB. On the left of this line class 1 points almost totally swamp class 2 points, and the converse is true on the right.

An approach to estimating resemblance that might seem obvious at first is to go straight for the negative of the variance of the estimated probabilities' $-V_k(\hat{f}(k|\hat{\mathbf{f}}(j|\mathbf{x})))$. The trouble is that this tells us how well the classifier *thinks* it separates the true classes, not how well it actually does separate them. Such a classifier could be hopelessly optimistic. For example, consider a situation which is quite inseparable (has a high, hence poor, value of some inseparability index) with all true probabilities $f(j|\mathbf{x})$ being nearly equal. Now suppose that the classifier's estimates of the probabilities are well spread out. Then $-V_k(\hat{f}(k|\hat{\mathbf{f}}(j|\mathbf{x})))$ will be low (a large negative value), which is good. But this will be a fairly useless indicator of any aspect of the classifier's performance; such a classifier will inevitably do little better than chance.

This measure, $-V_k(\hat{f}(k|\hat{\mathbf{f}}(j|\mathbf{x})))$, and similar ideas are essentially what was known as purity in Chapter 4, where they were used to decide how effective was a proposed split in constructing a tree. There the estimates $\hat{f}(k|\hat{\mathbf{f}}(j|\mathbf{x}))$ were obtained from the design set, rather than a test set. When *constructing* a tree we want the partitions to occur at positions where the classifier 'thinks' it produces good separation between the classes.

To overcome this problem for a measure of resemblance, we somehow have to estimate the $f(k|\hat{\mathbf{f}}(j\mathbf{x}))$, but not using the design set. This tells us how well differences in predicted probabilities *really* separate the classes; it is a genuine measure of resemblance. That means we need to study the differences between the *true* probabilities conditional on the $\hat{\mathbf{f}}(j|\mathbf{x})$, not the difference in *estimated* probabilities conditional on the $\hat{\mathbf{f}}(j|\mathbf{x})$, as was done by the impurity measure $-V_k(\hat{f}(k|\hat{\mathbf{f}}(j|\mathbf{x})))$.

A measure of resemblance thus conditions on the estimated probabilities and measures how different are the true probabilities associated with those estimated probabilities. In the case when the classifier divides the measurement space into regions within each of which $\hat{\mathbf{f}}(j|\mathbf{x})$ is estimated to be constant, such a measure looks at the partition induced by the classification rule and examines how the probabilities of true class membership in each region of this partition differ.

In general, separability is a measure of how well separated are the $f(j|\mathbf{x})$ (averaged over $\mathbf{x}$), resemblance is a measure of how well separated are the $f(j|\hat{\mathbf{f}}(j|\mathbf{x}))$ and impurity is a measure of how well separated are the $\hat{f}(j|\hat{\mathbf{f}}(j|\mathbf{x}))$. The problem is that f is unknown in these expressions. To produce a useful practical measure of resemblance we need to do something about the unknown true probabilities $f(j|\hat{\mathbf{f}}(j|\mathbf{x}))$. Equation (6.4) provides a way to overcome this: group the estimated probabilities according to ranges of $\hat{\mathbf{f}}(j|\mathbf{x})$ values then estimate the $f(j|\hat{\mathbf{f}}(j|\mathbf{x}))$ using the *test* set proportions in each class in each of these groups.

For the case of two-class grouped data, the decomposition in (6.4) becomes

$$D = \sum_{\alpha} v_{\alpha}\left[p_{\alpha}^*(1 - \hat{p}_{\alpha})^2 + (1 - p_{\alpha}^*)\hat{p}_{\alpha}^2\right] = \sum_{\alpha} v_{\alpha}(p_{\alpha}^* - \hat{p}_{\alpha})^2 + \sum_{\alpha} v_{\alpha}p_{\alpha}^*(1 - p_{\alpha}^*)$$

(6.5)

Example 6.3 gave an alternative decomposition which, for the case of two-class grouped data, becomes

$$D = \sum_{\alpha} v_{\alpha}\left[p_{\alpha}^*(1 - \hat{p}_{\alpha})^2 + (1 - p_{\alpha}^*)\hat{p}_{\alpha}^2\right] = \sum_{\alpha} v_{\alpha}(p_{\alpha}^* - \hat{p}_{\alpha})(1 - 2\hat{p}_{\alpha})$$
$$+ \sum_{\alpha} v_{\alpha}\hat{p}_{\alpha}^*(1 - \hat{p}_{\alpha}^*)$$

(6.6)

The first component can again be regarded as a measure of imprecision. Moreover, as we noted above, the second might be regarded as a measure of 'predicted resemblance'—how good is the rule, based on the assumption that it is precise. (Note that the second term of (6.6) involves the test set via the values of v_α.)

Decomposition (6.5) can only be computed from grouped data; otherwise the p_α^* take values of 0 or 1 and the inaccuracy and imprecision measures become the same. In contrast, (6.6) can be computed from ungrouped data. However, the imprecision term in (6.6) is not an ideal measure of imprecision; whatever the true imprecision, this measure will be small if the $(1 - 2\hat{p}_\alpha)$ factors are small (if $\hat{p}_\alpha$ is around 1/2 for all α, for example).

The second terms on the right-hand sides of (6.5) and (6.6) can be further decomposed as

$$\sum v_\alpha p_\alpha^*(1 - p_\alpha^*) = \bar{p}_\alpha^*(1 - \bar{p}_\alpha^*) - \sum v_\alpha(p_\alpha^* - \bar{p}_\alpha^*)^2$$

$$\sum v_\alpha \hat{p}_\alpha(1 - \hat{p}_\alpha) = \bar{\hat{p}}_\alpha(1 - \bar{\hat{p}}_\alpha) - \sum v_\alpha(\hat{p}_\alpha - \bar{\hat{p}}_\alpha)^2$$

where $\bar{p}^* = \sum v_\alpha p_\alpha^*$ and $\bar{\hat{p}} = \sum v_\alpha \hat{p}_\alpha$. In these decompositions the second terms on the right-hand side measure the variances of the p_α; large differences between the p_α mean good resemblance.

The logarithmic inaccuracy measure, $-\frac{1}{n}\sum_{i=1}^{n}\sum_j \delta(j|\mathbf{x}_i)\ln \hat{f}(j|\mathbf{x}_i)$, introduced in Section 6.3, with grouped data as above, becomes $-\frac{1}{n}\sum_{i=1}^{n}\sum_j f^*(j|\mathbf{x}_i)\ln \hat{f}(j|\mathbf{x}_i)$. Then, following the same procedure as above, we obtain the imprecision measure

$$-\sum_\alpha v_\alpha \sum_j f^*(j|\mathbf{x}_\alpha) \ln \hat{f}(j|\mathbf{x}_\alpha) + \sum_\alpha v_\alpha \sum_j \hat{f}(j|\mathbf{x}_\alpha) \ln \hat{f}(j|\mathbf{x}_\alpha)$$

$$= \sum_\alpha v_\alpha \sum_j f^*(j|\mathbf{x}_\alpha)\ln f^*(j|\mathbf{x}_\alpha)/\hat{f}(j|\mathbf{x}_\alpha)$$

which is the deviance again. Subtracting this from the above inaccuracy measure yields, after simplification $-\sum_\alpha v_\alpha \sum_j f^*(j|\mathbf{x}_\alpha) \ln f^*(j|\mathbf{x}_\alpha)$, which is entropy. Again it tells us how widely dispersed are the $f^*(j|\mathbf{x}_\alpha)$.

6.7   MEASURES OF INSEPARABILITY

Inseparability tells us how poorly separated are the true probabilities of belonging to each class. That is, it tells us how good a classifier built to distinguish between the classes could possibly be. At least, it *would* tell us this if we could calculate it; since we never actually know these probabilities, we never actually know the inseparability of the situation. We will need to estimate it. Inseparability itself, being solely a function of the actual distributions and their priors, is independent of any *estimated* probabilities or classifier rules, except insofar as they are based on a given set of variables. If the variable set, the measurement space, can vary then so can inseparability. Naturally, in order to enhance our chances of building a good classification rule, we will want to choose that variable set which yields the greatest separation between classes. So this is one important use of (estimates of) inseparability measures: they can be used to select sets of variables. Another is that they can indicate whether any rule we can construct (given a set of variables) is likely to be sufficiently effective for our purposes. (In the context of the literature on variable selection, 'separability measures' are usually used rather than inseparability. They are simply the converse of each other, with separability measures normally defined so that *higher values* are *good*. In our context, with lower values being consistently associated with good, it seemed more appropriate to adopt a different, complementary, term.)

To measure the inseparability between C distributions, we condition on $\mathbf{x}$, summarise how the probabilities of belonging to each of the C classes differ at that $\mathbf{x}$, then summarise that summary over the entire measurement space (the domain of $\mathbf{x}$). So let us focus on summarising how the probabilities of belonging to each of the C classes differ at some given $\mathbf{x}$. We shall call the measures conditional on $\mathbf{x}$ the *elements* of the inseparability measure. We can try to formulate a set of axioms which are attractive in such elements. The following axioms are common:

1. such elements should take their maximum when the classes have the same probability: $f(1|\mathbf{x}) = f(2|\mathbf{x}) = f(3|\mathbf{x}) = ...$ In this case, there is considerable inseparability (at $\mathbf{x}$) between the classes.
2. such elements should take their minimum when one class has probability 1 and the others have 0: $f(j|\mathbf{x}) = 1$ for some j. Then the separability cannot be greater.
3. such elements should be a symmetric function of their arguments $f(1|\mathbf{x})$, $f(2|\mathbf{x}), f(3|x), ...$

As an example for the two class case, we have used the element $\sum_j f(j|\mathbf{x})(1 - f(j|\mathbf{x}))$ yielding the separability measure $E_x \sum_j (1 - f(j|\mathbf{x}))f(j|\mathbf{x})$. This element satisfies the above three axioms. It can be generalised to more than two classes and can be rewritten as $\sum_j \sum_{k \neq j} f(j|\mathbf{x})f(k|\mathbf{x})$ or as $1 - \sum_j f(j|\mathbf{x})^2$ which is 1 minus the probability that a randomly selected and allocated point at $\mathbf{x}$ will be allocated to its correct class. This measure is sometimes called the *Gini index* (Chapter 4). Its relationship to the negative of the variance, which we proposed as a possible separability measure in Section 6.2, can be seen as follows. The negative of the variance is equal to $- \sum_j f(j|\mathbf{x})^2/C + \bar{f}^2$, where $\bar{f}$ is the mean of the $f(j|\mathbf{x})$ and therefore $\bar{f} = \sum_j f(j|\mathbf{x})/C = 1/C$ whatever the values of the $f(j|\mathbf{x})$ (since they sum to 1). That is, the negative of the variance is a simple linear transformation of $1 - \sum_j f(j|\mathbf{x})^2$.

Closely related measures are the *Chernoff measure*, for two classes taking the form $E\left\{ f(1|\mathbf{x})^s f(2|\mathbf{x})^{1-s} \right\}$ (with s between 0 and 1), and its special case, the *Bhattacharyya* measure, occurring when s is 1/2.

Another element satisfying the above axioms is $-\sum_j f(j|\mathbf{x}_i) \ln f(j|\mathbf{x}_i)$, or entropy.

An obvious approach is to look explicitly at the difference between the $f(j|\mathbf{x})$, as in the *Lissack–Fu* measure which uses the element $-|f(j|\mathbf{x}) - f(k|\mathbf{x})|^s$ and, for the two-class case, takes the form $-E |f(1|\mathbf{x}) - f(2|\mathbf{x})|^s$. The special case of this for $s = 1$ is called the *Kolmogorov variational distance*.

For the two-class case, the Lissack–Fu measure can be seen as a summary of the difference between the $f(\mathbf{x}, 1)$ and $f(\mathbf{x}, 2)$ distributions. Other such summaries can be easily derived, and they can lead to easily estimated measures. For example, several measures are based on summarising differences between objects, one taken from each of the (two) classes. Such measures are also used in cluster analysis (unsupervised pattern recognition). They include sums of squares of such distances, standardised sums of squares, and ratios of between-class to within-class sums of squares (as arise in multivariate analysis of variance test statistics). Some of them are equivalent to the negative of the variance measure we used above.

6.8 AN ALTERNATIVE APPROACH TO INSEPARABILITY

We have defined inseparability as a measure of how similar are the $f(j|\mathbf{x})$. The more distinct they are, the more confident one can be in one's classification (assuming that one has a classification rule which takes advantage of the differences). However, as we have stressed throughout this book, different applications lead to different problems. Consider a situation where the class conditional distributions $f(\mathbf{x}|j)$ are identical, but the priors are very different, with one class having a prior near 1. Then one can easily produce a classifier with a low error rate, simply classifying all objects to that class. The $\mathbf{x}$ values can add nothing to the accuracy of classification. Unfortunately, this may not always be adequate. (Indeed it seldom is. See the osteoporosis example in Chapter 8 and the discussion of screening in Section 8.2.) It may be essential to classify some points to each class. This can only be done in a rational way if there are differences between the $f(\mathbf{x}|j)$. And larger differences mean that more effective classifications can be produced. Thus we might define a new class of inseparability measures solely on the basis of differences between the $f(\mathbf{x}|j)$. (The class discussed above reduces to this case when the priors are equal.) Equivalently, we can summarise the differences between the $f(\mathbf{x}|j)$ and the overall mixture $f(\mathbf{x}) = \sum_j \pi_j f(\mathbf{x}|j)$.

Suitable measures are easy to invent, and several have been proposed. For two classes, the *divergence* or *information value* is

$$ - \int [f(\mathbf{x}|1) - f(\mathbf{x}|2)] \ln \frac{f(\mathbf{x}|1)}{f(\mathbf{x}|2)} \, d\mathbf{x} $$

generalising to the *Joshi* measure for more than two classes

$$-\sum_j \pi_j \int [f(\mathbf{x}|j) - f(\mathbf{x})] \ln \frac{f(\mathbf{x}|j)}{f(\mathbf{x})} \, d\mathbf{x}$$

and the *Matusita* measure

$$-\sum_j \pi_j \left\{ \int \left[\sqrt{f(\mathbf{x}|j)} - \sqrt{f(\mathbf{x})} \right]^2 d\mathbf{x} \right\}^{1/2}$$

Earlier we measured inseparability by conditioning on **x**, measuring the local inseparability at **x** and summarising over the measurement space of the **x**s. Put another way, we summarised differences to yield an overall measure. An alternative approach is to compute the difference of summaries. In just about the simplest example one can imagine, we may summarise each of the distributions $f(\mathbf{x}|1)$ and $f(\mathbf{x}|2)$ by their means and compute the difference between the means. A large difference would imply great separability and a low inseparability. (In keeping with the conventions we have adopted about what is good and what is bad, we would need to define inseparability as the negative of the absolute difference between the means.)

Several such measures have been based on this approach. For example, with two classes we have the following measures:

- Minkowski metrics, $-\left[\sum_i |\mu_{1i} - \mu_{2i}|^s\right]^{1/s}$, where μ_{ji} is the ith component of the mean vector of class j.
- Generalisations of the Mahalanobis distance, $-(\mu_{1i} - \mu_{2i})^T A (\mu_{1i} - \mu_{2i})$, where A is some symmetric matrix (such as the average covariance matrix of the two classes).

6.9 CONFUSION MATRICES

Confusion matrices are normally expressed using error rate, but the concept can be generalised. When the performance criterion is error rate, a confusion matrix is the cross-classification of the predicted class by the true class. The population matrix will have probabilities in each of the cells, and one can define an estimate of it in terms of test set proportions falling in each cell. (One can also define a matrix based on the design set, with the usual provisos about it being optimistically biased as an estimate of the true cell probabilities.)

When there are only a few or a moderate number of classes, the confusion matrix is a convenient way of summarising classifier performance. The off-diagonal elements show where the main misclassifications occur. Denoting an off-diagonal probability by p_{ij} and the corresponding marginal probabilities by p_i (for the predicted proportion falling in class i) and π_j (for the true proportion in class j), the product $p_i \pi_j$ shows the proportion of objects we would expect to find in the ijth cell if the classification were made randomly, subject to forcing the class sizes to be of the given size. In particular, $p_i p_j$ is the proportion we would expect to find in the ijth cell if we insisted that the predicted class sizes were the same as the true class sizes. The difference $p_{ij} - p_i p_j$ shows how well the classifier would perform relative to random assignment (and one could identify particular cells where this would matter). This matrix of 'residuals' may be more useful than a simple examination of the raw confusion matrix. Measures of performance relative to chance assignment can be based on chi-squared statistics.

The confusion matrix, based on error rate, is asymmetric because error rate itself is a fundamentally asymmetric measure; there is no reason why the number of objects from class i misclassified into class j should equal the number from class j misclassified into class i. But we have also discussed symmetric measures, which would lead to symmetric confusion matrices; this is a generalisation of the term *confusion matrix*—it usually refers to the error rate matrix. In general, any measure of the distance between two classes can be used as the basis of confusion matrices.

6.10   A NOTE ON TERMINOLOGY

The terms *inaccuracy, imprecision, inseparability* and *resemblance*, defined in Section 6.2, are nonstandard. Here we describe why we have coined these terms and what alternatives are found in the literature.

We have explained that *low* values of all four concepts are *good* in terms of classifier performance and we wanted terms which reflected this semantic content. So, for example, a low value of our inaccuracy measure means that the classifier is a good one; by implication such a classifier has a high value of accuracy. Similarly, a low imprecision value means a high precision (the rule's estimates are precise). Low inseparability (high separability) is desirable and is associated with one class having a higher membership probability at each $\mathbf{x}$. Finally, low resemblance means that the probabilities conditional on $\hat{f}(j|\mathbf{x})$ are very different (perhaps one may dominate). Our concept of *impurity*, discussed in Chapter 4, also satisfies this convention.

In place of *imprecision*, Miller (1962) uses *validity*, Murphy (1973) uses *reliability*, and Dawid (1982) uses *calibration*. To us it seemed that these terms have an awkward semantic ordering; a reliable rule might be expected to be one which had a large score on a measure of reliability. Certainly, this is the usage in other contexts (e.g. Dunn, 1989).

In the place of *resemblance*, Sanders (1963, 1973) uses the term *resolution* and DeGroot and Fienberg (1983) use the term *refinement*. Again, we hope to produce rules which have low values for this measure; a low resemblance between probabilities of belonging to each of the classes at $\mathbf{x}$ is to be hoped for.

Hilden *et al.* (1978b) uses the term *discriminatory ability* for *inaccuracy*, *reliability* for *imprecision*, and *sharpness* for the extent to which a rule assigns a large probability to just one class, a measure of how distinct are the $\hat{f}(j|\mathbf{x})$. Hand (1994c) uses the terms *discriminability, separability* and *reliability* for *inaccuracy, inseparability* and *imprecision*, respectively, and Hand (1995a) extends them to adopt *refinement* for *resemblance*. Again, the semantics of discriminability (inaccuracy) suggest that a good classification rule might be expected to be one which scores high on a measure of 'discriminability', signifying large discriminating power, in contrast to our measures, where low is good. And in a similar way classification problems generally aim for a *high* degree of *separability* between classes in the measurement space; this means that a decision surface can accurately distinguish between them. Thus to use separability in place of *inseparability* (which we want to take low values) would lead to awkwardness.

While on the subject of terminology, Yates (1982) refers to the general 'extent to which probabilistic forecasts do anticipate the events at issue' as *external correspondence*. In the place of inaccuracy, we have already noted that measures of discriminability are sometimes called *scoring rules, quasi-utility* functions or *pseudo-utility* functions, but these are general terms describing such measures rather than their values.

We have already noted above that *separability* measures, used in the context of variable selection, play the same function as our inseparability measures, but are typically defined so that large separability (between distributions) is good. Since our measures naturally associate low values with good rules, it seemed sensible to adopt a different terminology. However, separability measures and inseparability measures play the same roles, although from complementary perspectives.

6.11 FURTHER READING

In the context of Bayesian forecasting, various decompositions of inaccuracy are given in Yates (1982) and an elegant discussion of one particular decomposition is given in DeGroot and Fienberg (1983). Amongst the earliest work in this context are three articles by Murphy (1972a, 1972b, 1973).

Habbema *et al.* (1978) and Hilden *et al.* (1978a, 1978b) discuss a variety of measures of inaccuracy, imprecision and resemblance, and Hand (1994c) extends the discussion to inseparability. Hand (1995a) links this material into the Bayesian forecasting work.

Breiman *et al.* (1984, Chapter 4) discuss impurity measures in the context of decision trees (see Section 4.3). Aczel and Daroczy (1975) present axioms for measures of information and generalised entropy. The particular measures mentioned throughout this chapter are described in Bhattacharyya (1943), Chernoff (1952), Matusita (1955), Joshi (1964) and Lissack and Fu (1976). Divergence is discussed in Jeffreys (1946) and Kullback (1959). Further discussion of separability measures from a pattern recognition perspective is given in Devijver and Kittler (1982).

Multivariate analysis of variance test statistics are discussed in Hand and Taylor (1987, Chapter 4) and Hand and Crowder (1996). Krzanowski (1983a) has reviewed separability measures, classing them as either based on ideas from information theory or related to Bhattacharyya's measure. Krzanowski (1983a, 1984, 1987) examined Matusita's distance with mixed continous and categorical variables.

CHAPTER 7

Misclassification Rate

7.1 INTRODUCTION

The *misclassification rate* or *error rate* of a classifier is the proportion of objects which it misclassifies. Normally, this will refer to future objects to be classified by the rule, so that the complement of error rate tells us what proportion of future objects we can expect to classify correctly. Error rate is by far the most popular criterion for assessing the performance of an allocation rule and yet, as described in Section 6.1, it is not without disadvantages. For example, it does not accord different costs to the different kinds of misclassification and it does not penalise errors which result from estimated probabilities far from true ones. On the other hand, if rules are to be compared then some criterion has to be used for the comparison; and error rate does have some attractive features, especially its simplicity and ease of understanding.

In assessing the merits of error rate as an assessment criterion, it is useful to distinguish between applications and methodological investigations. In applications, with a particular problem in mind, some other criterion may be more natural. In the credit scoring example of Section 10.3, a natural criterion is the proportion of accepted applicants who in fact turn out to be bad risks, given that a specified proportion is accepted, and in medical applications one might be interested in specificity for a given sensitivity. Even in real applications it is common for the criterion to be loosely defined—the stated objective may simply be to identify the 'better' rule. Error rate might then be adopted, but some careful analysis of the problem and its objectives should be undertaken first.

In methodological studies, for example in simulation studies, it is perhaps even more common for there to be no specific criterion defined at all. One might simply be seeking to identify which rule is 'better' in a general sense, so as to guide likely future users. Then it might be as well to adopt several criteria, perhaps seeking to make them complementary to some extent. Given the popularity of error rate, one would presumably want to include this particular criterion.

Since error rate is so popular, it is hardly surprising that considerable research effort has been put into finding accurate estimators of it. Indeed, Toussaint (1974) commented that error rate estimation was one of the most important problems in pattern recognition. Of course, superficially, estimating error rate is straightforward: one simply sees what proportion of cases are misclassified. However, this description conceals difficulties; for example it is well known that simply reclassifying the design set and counting the proportion of misclassifications will typically lead to an underestimate of future error rate. Moreover, there are subtleties of definition about precisely which error rate is being estimated; the term *error rate* has more than one meaning. In view of this, we use Section 7.2 to clarify the different kinds of error rate.

7.2   TYPES OF ERROR RATE

The *Bayes error rate* is the minimum possible error rate given a set of measurements. It is the error rate which would result given complete knowledge of the class conditional probability distributions. In a word, it is the *optimal* error rate. It provides a lower bound on any error rate which may be achieved by a real classification rule. Letting $f(\mathbf{x})$ represent the overall distribution of measurement vectors $\mathbf{x}$ and $f(j|\mathbf{x})$ the probability of belonging to class j at $\mathbf{x}$, the Bayes error rate is

$$e_B = \int \left[1 - max_j f(j|\mathbf{x})\right] f(\mathbf{x}) d\mathbf{x}$$

A related form of error rate is the *best achievable* error rate e_b for a given form of classifier. Suppose that the data in fact arise from multivariate normal distributions with unequal covariance matrices, so that the optimal decision surface is quadratic. This will be unknown to the investigators and they might try using a linear classifier. Then e_b is the smallest error rate which can be obtained from such a classifier. In general, of course, $e_b \geq e_B$. e_b is useful because it tells us how far our rule is from achieving the best that a classification rule from the chosen family of rules can achieve. From this we might decide to accept the rule we have, to attempt to improve it while retaining the same family, or to choose a different family.

Any given classification rule merely estimates the $f(j|\mathbf{x})$ (or some transformation of them) and, on the basis of these estimates, defines regions R_j in which the rule classifies points to class j. The *actual* error rate for such a classifier is then

$$e_C = \sum_j \int_{Rj} [1 - f(j|\mathbf{x})] f(\mathbf{x}) d\mathbf{x}$$

Some authors call this the *true* error rate, some the *conditional* error rate because it is conditional on the design set used in building the classifier. We can also think of it as the error rate which would be obtained if the classifier were applied to an infinite test set from the same distribution as the data used in its construction. The difference between e_B and e_C arises from the fact that the design set is merely a sample.

From this one can define the *unconditional* or *expected* error rate e_E, which is the expected value of e_C over design sets of the given size. In practice one must also decide if one intends to condition on the relative numbers falling in each class.

Given a design set (in a particular application), one will be interested in the conditional error rates e_C of the rules. However, if one is in the situation of having to decide what rules to adopt before seeing the data, one will be interested in the unconditional rate e_E. Put another way, if one wants to compare *estimators*, one will use e_E, but if one wants to compare *estimates* one will use e_C. More rarely will one be interested in e_B and e_b; most research on error rate estimation has focused on e_C.

In what follows, we assume that one is interested in overall error rate (combined over classes). If, however, one is interested in individual rates—the proportion of class 1 objects which are misclassified, for example—then corresponding estimates can be derived by focusing on the classification of points from each class individually.

7.3 ESTIMATING THE ACTUAL ERROR RATE

An obvious initial approach to estimating actual error rate is to reclassify the design set and determine the proportion of objects misclassified. Unfortunately, this will typically lead to an optimistic (under)estimate of the error rate on future cases. This is because the rule will have been optimised, in some sense, for performance on the training data; error rate or some related performance measure will have been minimised, say, relative to the design data. Future objects, not possessing the distributional idiosyncrasies of the design sample, will have a slightly different distribution and hence should be expected to have a larger value of error rate. This property is clearly related to issues of overfitting, discussed elsewhere in the book. It follows that the problem will be less severe the larger is the design set. Asymptotically the design and test distributions will be the same. It also follows that the problem will be less severe the less flexible is the classifier. This leads to the compromise between oversmoothing and overfitting discussed in Chapter 1. The estimate of error rate obtained by reusing the design data is termed the *resubstitution* or *apparent* error rate.

Having said all that, it is perhaps worth remarking that nowadays, as a consequence of progress in automated data acquisition techniques, more and more very large data sets are being analysed. They arise, for example, in marketing and commercial applications from electronic point-of-sale data capture, as well as in areas such as character recognition and image analysis. With vast data sets the scope for overfitting the design set is often negligible, even with very flexible classifiers, so one typically finds that the resubstitution estimate is very similar to the estimate obtained using a test set, and the sophisticated error rate estimation methods described below are not needed.

For applications without huge data sets, the simplest and most obvious way of estimating actual error rate, without incurring the resubstitution bias, is to count the proportion of objects which the rule misclassifies in a *test set* of objects independent of those in the design set. We shall call this the *error count estimate*. This estimator has the merit of being unbiased and easy to calculate. Care must be taken to ensure that the test sample really has been drawn from the same distribution as the design sample and is not, for example, distorted in some way by the process of selection or division of a large sample into design and test sets. A common cause of such bias is population drift, where the population changes over time, especially between the design of the rule and its application or evaluation. This can be a particular danger in commercial applications, where the population being classified consists of customers, who are vulnerable to changes in the overall economic environment. More generally, the test data must reflect the population to which the classifier is to be applied (whether or not this has precisely the same distribution as that from which the design data were drawn), in terms of both distributional forms and prior probabilities.

Although simple, this independent test set approach is not universally applicable. If a test set is available at the time of construction of the classifier one is led to ask why it has not been used in the construction—the larger design data set would yield a better rule. More critically, if the complete data set is not very large, one will be loath to divide it and use only part for design. Also, if a data set is to be partitioned into a design and test set, one has to decide into what proportions to divide it.

A compromise between splitting the data into independent design and test sets, with its consequent sample size reduction, and simply designing using all the data is *cross-validation*. This involves extracting subsets of the data to test the performance of classifiers built using the remainder, then repeating this for different subsets and averaging the results. Cross-validation has the following variants:

- *Rotation*: say 10 mutually exclusive subsets are defined, each one being used in turn as a test set for the classifier built on the remainder.
- *Leave-one-out*: a single point is used as a test set for the classifier built on the other $n - 1$; this is repeated n times.
- *Bootstrap*: a random subset of size equal to the complete data set is taken, with replacement, for use as the design set, and the complete data set is used as the test set.

There are many variants of the bootstrap method and we outline some of them below. Bootstrap methods have had a big impact on statistics as a whole. Their early genesis seems to have been in problems of error rate estimation, demonstrating yet again how classification problems are a paradigmatic statistical problem.

With all methods which involve splitting the available data into design and test sets, care must be taken to ensure that the class sizes are handled appropriately. Reasonable estimates for the class priors will be the class sizes in the complete data set (if this is a random sample from the overall population). Presumably, one will usually use these estimates in conjunction with the decision surface estimates obtained from the design subsample to classify the test points. Similarly, the test set objects in each class will yield an estimated misclassification rate for that class. These rates should be combined using appropriate priors (and presumably the priors will usually be the class sizes in the complete data set) rather than the class sizes which happen to occur in the test data.

The *jackknife* method, superficially similar to the leave-one-out method, is in fact based on rather different ideas. We describe it below.

Finally, an entirely different class of estimators is the *smoothing* or *average conditional error rate* approach, also discussed below.

7.4   THE LEAVE-ONE-OUT METHOD

Sometimes the term *cross-validation* is used to describe the particular leave-one-out form. The advantage of this extreme approach is that, in each case, the design set is almost as large as the entire data set. Since, presumably, the classifier which will be used in practice will be that based on the complete data set, this means that the estimate of error rate is approximately unbiased (any bias arises solely from the extra variation in the position of the decision surface due to using a design set of size $n-1$ instead of n; it means that the expected performance of each of the n classifiers will be very slightly worse than the performance of the classifier based on all n objects).

The method is elegant, but there is evidence that it has a relatively large variance. Moreover, in general, such an approach will obviously require extensive computing resources, since n separate classifiers must be built. Having said that, there are special cases in which properties of the classification rule estimators can be used to develop updating procedures which sidestep the need for recalculation from scratch (most notably, methods based on linear models and least squares, such as classical discriminant analysis). In any case, progress in computer technology means that such problems become less serious with time, and assessment of future error rate is rarely a real-time problem.

7.5   BOOTSTRAP METHODS

Cross-validation attempts to remove the apparent error rate bias by splitting the available data so that a test set independent of the design set can be used. An alternative strategy for removing the bias is to estimate it then to use this estimate for adjusting a biased estimator. Bootstrap methods and jackknife methods described in the next section follow this approach. For a fixed design set the bias is $e_C - e_A$, with e_C the true error rate and e_A the apparent error rate. Now the empirical distribution $\hat{f}(\mathbf{x})$, taking value $1/n$ at each $\mathbf{x}$ and 0 elsewhere, can act as an estimate of the true distribution $f(\mathbf{x})$. In fact, this is how it is normally used; we estimate population values by summarising $\hat{f}(\mathbf{x})$. In our case we will compute an estimate of $e_C - e_A$ from $\hat{f}(\mathbf{x})$. This is straightforward to do. $\hat{f}(\mathbf{x})$ is obtained from $f(\mathbf{x})$ by taking a sample, so we take a subsample $\hat{f}_k(\mathbf{x})$ of the same size n (and hence with replacement) from $\hat{f}(\mathbf{x})$. The apparent error rate of this subsample e_{Ak} can be computed directly. Moreover, and this is the essence of the idea, the true error rate of this subsample e_{Ck}, viewed as a sample from the 'population' distribution $\hat{f}(\mathbf{x})$ can be computed (simply by classifying *all* of the objects in the sample, not merely those in the subsample). The difference $e_{Ck} - e_{Ak}$ estimates $e_C - e_A$. Of course, the size of $e_{Ck} - e_{Ak}$ depends upon the particular subsample drawn, and so is variable. We overcome this by drawing multiple bootstrap samples ($k = 1, ..., m$) and averaging the results. The final bootstrap error rate estimate e_β is obtained by adding the estimate of bias to the apparent error rate obtained from the full sample.

This basic bootstrap estimator has been extended in a number of ways, some of which we now briefly outline.

The above estimate of error rate has the form $e' + (\hat{e} - e')$, where e' is an observable statistic calculated from the data. The estimate is then obtained by estimating $(\hat{e} - e')$ by its bootstrap expectation. Put like this, there is no reason why e' should not be replaced by any other observable statistic. One suggestion following these lines is to replace e' by e_β, the original basic bootstrap estimator. This yields the *double bootstrap*.

A weakness of using $\hat{f}(\mathbf{x})$ as an estimate of $f(\mathbf{x})$ is that $\hat{f}(\mathbf{x})$ has relatively few possible values, whereas $f(\mathbf{x})$ may be continuous. Consequently, for a randomly chosen sample member, there is a high probability that it will be identical to a member of the bootstrap subsample used to classify it, which is quite different from the true situation. This probability is $1 - (1 - 1/n)^n$, which tends, in the limit as sample size increases, to $(1 - 1/e) \approx 0.632$. From this it has been shown that an improved estimate of error rate is given by $0.368e_A + 0.632e'_\beta$, where e'_β is the bootstrap error rate for sample points which are *not* included in the bootstrap subsample classifying them. This estimator is termed the *632 bootstrap* and empirical studies suggest it is one of the best that has been developed. In fact it turns out that e'_β can be well approximated by a cross-validation approach based on using half the data for the design set and half for the test set, doing this repeatedly and averaging the results.

The idea of *shrinking* (which reduces variance but perhaps at the cost of increasing bias, as in the regularisation techniques discussed in Section 2.7) might also be used to lead to improved estimators. In the two-class case, the empirical distribution puts all of its weight on $(c_i, \mathbf{x}_i)$, where c_i is the true class of the ith point $\mathbf{x}_i$. The *randomised bootstrap* instead draws its bootstrap samples from a distribution in which a nonzero probability is assigned to both $(c_i, \mathbf{x}_i)$ and its complementary point $(\tilde{c}_i, \mathbf{x}_i)$. For example, the probability 0.9 might be assigned to $(c_i, \mathbf{x}_i)$ and 0.1 to $(\tilde{c}_i, \mathbf{x}_i)$. This method can also be viewed as a smoothing approach, though here the smoothing is entirely in the 'class membership direction'.

7.6 THE JACKKNIFE

Another technique for attempting to remove the bias of the apparent error rate is the *jackknife* method. This is computationally less demanding than the bootstrap, since it merely involves combining values derived from n subsamples, each one being the original design set with one element removed (like leave-one-out, although the rationale is different). The jackknife was developed before the bootstrap, but it turns out to be an approximation to the bootstrap.

Let e_A be the apparent error rate, let e_{Ai} be the apparent error rate based on the sample with the ith observation removed, and let $e_{A\bullet} = \frac{1}{n}\sum_i e_{Ai}$, the average of the n apparent error rates based on samples of size $(n - 1)$. Then the jackknife estimate of bias is $(n - 1)(e_A - e_{A\bullet})$. Using this to adjust the overall apparent error rate, we obtain the estimate $e_J = e_A + (n - 1)(e_A - e_{A\bullet})$. This has bias of $O(n^{-1})$ as an estimator of the *conditional* error rate but bias of $O(n^{-2})$ as an estimator of the *unconditional* error rate. To produce an estimator of the conditional error rate with bias of $O(n^{-2})$ replace e_A in $(n - 1)(e_A - e_{A\bullet})$ by $e_{A(\bullet)=\frac{1}{n}\sum_i e_{A(i)}}$, where $e_{A(i)}$ is the proportion of the entire design sample misclassified by a design set consisting of the entire sample minus the ith point. Hence $e_{J\bullet} = e_A + (n - 1)(e_{A(\bullet)} - e_{A\bullet})$.

The relationship between the jackknife and leave-one-out estimators can be seen as follows. A little new notation is needed. Let $e(i)$ be 0 if the ith object is correctly classified by the rule and 1 otherwise. Let $e_{(j)}(i)$ be 0 if the ith object is correctly classified by a rule based on the design set with the jth object removed, and 1 otherwise. Now, the jackknife estimator is

$$e_{J\bullet} = e_A + (n - 1)\left(e_{A(\bullet)} - e_{A\bullet}\right)$$

$$= \frac{1}{n}\sum_i e(i) + (n - 1)\left[\frac{1}{n^2}\sum_i\sum_j e_{(i)}(j) - \frac{1}{n(n-1)}\sum_i\sum_{j\neq i} e_{(i)}(j)\right]$$

$$= \frac{1}{n}\sum_i e(i) + (n - 1)\left[\frac{1}{n^2}\sum_i\sum_j e_{(i)}(j) - \frac{1}{n(n-1)}\left\{\sum_i\sum_j e_{(i)}(j) - \sum_i e_{(i)}(i)\right\}\right]$$

$$= \frac{1}{n}\sum_i e(i) + \frac{1}{n}\sum_i e_{(i)}(i) - \frac{1}{n^2}\sum_i\sum_j e_{(i)}(j)$$

$$= e_L + e_A - e_{A(\bullet)}$$

where e_L is the leave-one-out estimate.

7.7 AVERAGE CONDITIONAL ERROR RATE METHODS

We can write the true error rate of a rule as $e_C = \int e(\mathbf{x}) f(\mathbf{x}) d\mathbf{x}$, where $e(\mathbf{x})$ is the conditional probability of error at $\mathbf{x}$ and $f(\mathbf{x})$ is the overall mixture distribution. Decomposing e_C in this way allows us to focus on the two components separately. Note in particular that estimates of $f(\mathbf{x})$ can be obtained without knowledge of the true classes of the test points. This can be valuable if the test data are easy to obtain but the true classes are awkward to determine because the process is time-consuming, expensive or suffers from some other drawback. Because of the decomposition, such estimators are called *average conditional error rate* estimators.

If we put $\hat{f}(\mathbf{x})$ equal to $1/n$ at the n test points and zero elsewhere, and define $\delta(c_i, \hat{c}_i)$ to be 1 if the true class c_i equals the predicted class $\hat{c}_i$ and zero otherwise, it yields the estimate $\hat{e}_C = \frac{1}{n} \sum_i \delta(c_i, \hat{c}_i)$. This is the simple error count estimate if the n points are an independent test set, and the apparent error rate if they are the design set. In this, the $\delta(c_i, \hat{c}_i)$ can be viewed as providing estimates of the local error rate, so that the result is indeed an average conditional error rate estimator. Of course, the $\delta(c_i, \hat{c}_i)$ are the most unsmooth of these local estimates; in a sense they are those estimators which yield the maximum variance locally (although they are unbiased). This variance can be reduced by shrinking—replacing the $\delta(c_i, \hat{c}_i)$ by alternative values shrunk away from 0 and 1. For example, points close to the decision surface can have $\delta(c_i, \hat{c}_i)$ shrunk strongly towards 0.5 and points a long way from the decision surface shrunk hardly at all. (Bias introduced by the smoothing can be reduced by bootstrapping.) If the local error rates are estimated by the k-nearest neighbour method then, asymptotically, the variance of the smoothed error count estimator satisfies

$$v(\hat{e}_c) \le \frac{e_C(1 - e_C)}{n} - \frac{(k - 1)(1 - \mu)e_C}{kn}$$

where e_C is the true error rate and $\mu = max_x e(x)$. Since the first term on the right-hand side is the variance of the error count estimator, this shows that smoothing does reduce variance, at least asymptotically.

7.8 OTHER POINTS

The question to be answered, of course, is, Which method should one choose? Up until about 1970 apparent error rate seemed to be common but then cross-validation became ascendant, especially the leave-one-out variant. However, the relatively large variance of this method has now driven it out of favour and bootstrap methods seem to be preferred, with the 632 being apparently the current method of choice.

We noted above that the distinction between conditional (actual) and unconditional error rates has implications in comparisons between rules. In a real problem, presented with a real finite sample, one will be interested in the conditional rates. Conversely, unconditional error rate is of interest for methodological work where one is trying to establish which rule is better in the sense that it will be *expected* to have a smaller error rate in future (real) applications. The majority of methodological studies seem to compare rules using simulation: they average the results over many generated design sets, so evaluating unconditional rate.

Fitzmaurice *et al.* (1991) noted that it may not always be appropriate to regard, say, the difference between error rates of 0.03 and 0.01 as equally important as the difference between, say, 0.23 and 0.21. Although in both cases the difference is 0.02, it is likely that the former will be of much greater significance. This can be reflected in comparative studies by transforming the error rate results before combining them, perhaps by a logistic transformation.

Another point, raised by Hand (1987a) and pursued by Fitzmaurice *et al.* (1991) is that the definition of bias (and hence of mean square error) in simulation studies in this area often differs slightly from the conventional definition. For given population distributions (from which the data are generated) the design samples will have true error rates which differ from sample to sample. 'Bias' in such studies is often computed as an average over simulations of (estimated error rate − true error rate) for the generated data set, and here 'true error rate' varies. Such results are typically grouped together, however, in terms of the Bayes error rate, thus averaging over different true error rates.

Occasionally one may be interested in estimating the Bayes error rate, the best rate that can be achieved with a given set of variables. In fact the last phrase indicates why such a measure might be useful; it can be used to distinguish between variable sets to identify a set which will permit sufficiently accurate classifications. More generally, one might simply want to know whether there is any hope of adequate classification accuracy for a given set of variables. One approach to estimating the Bayes error rate is based on the fact that the error rates of nearest neighbour methods provide asymptotic bounds on the Bayes error rate, as mentioned in Section 5.3.

We should also mention the reject option in the context of error rate estimation. As has been noted before, this is the term used to describe the strategy of rejecting some objects from the classification process if one cannot be sufficiently confident about their class. Thus, one might choose only to classify those objects for which $\max_j \hat{f}(j|\mathbf{x})$ is greater than some threshold. By this process one can make the proportion correctly classified arbitrarily high, though perhaps at the expense of classifying only a small proportion of the population. Depending on the objectives of the problem, this may be acceptable. In marketing, for example, one may want to be very confident that the accepted customers will be profitable, and it will not matter that a large proportion of potential customers, including many who would have been profitable, have been rejected. In other problems, such as medical diagnosis, it is unacceptable simply to reject subjects from the classification process. The idea can still be used but, instead of rejecting subjects, one defers a decision while more information is collected (more variables measured). This leads to the sequential classification process outlined elsewhere. Of course, there is no guarantee that by collecting further information one can ultimately reduce the error rate to an acceptable level; it will depend on the size of the available design set.

The reject option has been explored in considerable depth in the context of error rate with nearest neighbour methods, but the principle can be applied generally with any performance measure and any classification rule, only including those regions of the measurement space for which one is very confident of the classification.

7.9 FURTHER READING

Reviews of error rate estimation methods are given by Toussaint (1974), Hand (1986), McLachlan (1987) and McLachlan (1992, Chapter 10). Piper (1992) provides a nice illustration of how test set and apparent error rate approach each other as the size of the data sets increase (using some very large chromosome data sets). Figure 1 in that paper shows (i) test set error rates decreasing asymptotically to the true error rate with increasing sample size and (ii) apparent error rates increasing asymptotically to the true error rate with increasing sample size. The two curves are almost exact mirror images.

The resubstitution estimate, or apparent error rate, was proposed by Smith (1947) and the first formal description of the leave-one-out method seems to be that of Lachenbruch (1965). An accessible description of the down-dating strategy to avoid repeating all the calculations in the leave-one-out method is given in Lachenbruch (1975, p.36).

The idea of the jackknife was originally proposed by Quenouille (1949). The jackknife method is discussed in Efron (1982) and Efron and Gong (1983). Efron (1982) and Efron and Tibshirani (1993) show relationships between jackknife, bootstrap and leave-one-out methods. In particular, Efron and Tibshirani (1993) is a comprehensive review of bootstrap and related methods.

The formal development of the 632 bootstrap, as summarised above, is due to Efron (1983), which also describes the other bootstrap variants. However, it had earlier been described by Toussaint (1975) and McLachlan (1977), before the advent of bootstrap methods, who based it on the observation that, since e_A was optimistically biased and e'_β pessimistically biased, perhaps a suitable linear combination could be formed which was essentially unbiased. Toussaint relates the 632 method (though he does not use that term) to even earlier work based on this linear combination principle. A more recent Monte Carlo comparison of the 632 estimator with others is presented in Fitzmaurice et al. (1991).

Fukunaga and Kessell (1973), Kittler and Devijver (1982) and Fitzmaurice and Hand (1987) estimated $e(\mathbf{x})$ in average conditional error rate estimators by local smoothing—kernel or k-nearest neighbour methods. In contrast, Glick

(1978), who used the term *additive estimators* for what we have called average conditional error rate estimators, suggested an overall model for $e(\mathbf{x})$, namely the logistic function. Other authors who have used global models include Ganesalingam and McLachlan (1980) and Snapinn and Knoke (1985, 1988, 1989).

Work on estimating the Bayes error rate includes that by Fukunaga and Kessell (1973) and Garnett and Yau (1977).

Krzanowski and Hand (1996) have explored the implications of the fact that, in many simulations studies, the true error rate varies from case to case.

The reject option has been explored by Anderson (1969), Habbema, *et al.* (1974) and, especially for nearest neighbour methods, by Devijver and Kittler (1982).

CHAPTER 8

Evaluating Two-Class Rules

8.1 CRITERIA

The two-class situation is the most common special case, and in this section we examine the assessment of two-class classification rules in detail. One might suppose that, being also the simplest special case, with a confusion matrix which has only four cells, there would not be much to say about it. However, this turns out not to be true: the reason that it is so common is that it arises in many different situations and each such situation places emphasis on a different aspect of it.

The general notation we shall use is presented in Table 8.1. Here π_0 is the prior probability of class 0, $\pi_1 = 1 - \pi_0$, p_0 is the proportion predicted to have come from class 0, $p_1 = 1 - p_0$, and n is the overall sample size $(= a + b + c + d)$.

To measure performance based on Table 8.1, and subsequently to use such measures to optimise and/or choose between classifiers, we need to reduce it to a single summary statistic. For a given data set, n and $a + c$ (and consequently $b + d = n - a - c$) are fixed, so reducing the four numbers to two degrees of

Table 8.1 Confusion matrix notation for the two-class case.

		True class		
		0	1	
	0	a	b	p_0
Predicted class	1	c	d	p_1
		π_0	π_1	n

freedom. However, there is no necessary constraint which will reduce these two to one. We have to choose or invent a constraint to do this. Before looking at ways in which this might be achieved, we examine some approaches which present results for two degrees of freedom, letting the user choose between classifiers.

A strategy common in epidemiology is to report the *sensitivity* (*Se*) and *specificity* (*Sp*). If class 0 represents 'cases' and class 1 represents 'noncases', these are defined respectively as $a/(a + c)$ and $d/(b + d)$. Sometimes sensitivity and specificity are called *true positive rate* and *true negative rate*, respectively. Hence, as an alternative, *false negative rate* $(c/a + c)$ and *false positive rate* $(b/b + d)$ are sometimes given. Sensitivity and specificity define performance in terms of predicted classifications within each of the true classes. Complementary to this, defining performance as proportions correct within those predicted to belong to each class, we have the *positive predicted value* $(a/a + b)$ and the *negative predicted value* $(d/c + d)$. Obviously the two 'predicted values' can be derived from sensitivity and specificity and vice versa, given that only two degrees of freedom are involved (given that $a + c$ and $b + d$ are fixed).

Often the two degrees of freedom are presented simultaneously for a range of possible classification thresholds for the classifier in a *receiver operating characteristic (ROC) curve*. This is done by plotting true positive rate (sensitivity) on

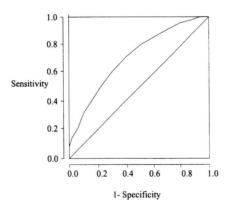

Figure 8.1   An example of an ROC curve.

the vertical axis against false positive rate (1−specificity) on the horizontal axis (equivalent variants are sometimes used). An example is given in Figure 8.1, which shows the ROC curve for a nearest neighbour classifier on a data set of applications for home improvement loans. Different points on the curve correspond to different thresholds used in the classifier. Diagrams like Figure 8.1 are sometimes called *Lorentz diagrams*.

A classifier with an ROC curve which followed the 45° line would be useless. It would classify the same proportion of the cases and non-cases into the case class at each value of the threshold; it would not separate the classes at all. In contrast, a perfect ROC curve would follow the two axes. It would classify 100% of the cases into the case class and 0% of the noncases into the case class for some value of the threshold. Real-life classification rules produce ROC curves which lie between these two extremes with, in general, better rules producing curves nearer to the optimum curve.

By comparing ROC curves one can study dominance relationships between classifiers. For a given specificity the curve which has greater sensitivity will be superior. Similarly, for a given sensitivity, the curve which has greater specificity (therefore smaller false positive rate) will be better. If and only if two curves touch or cross at a point will they give identical performance (at the threshold values corresponding to the point of intersection). All ROC curves intersect at the bottom left and top right corners of the square. Sometimes the curve for one classifier dominates (is superior to) that for another at all values of the threshold; one curve is higher than the other throughout the diagram. The more common situation, however, is that one curve dominates in some interval of thresholds and another dominates in other intervals. Then one can look to see if either curve dominates within the range which is thought to be realistic for the problem at hand.

If one knows that misclassifying a class 0 object incurs a cost of c_0 and misclassifying a class 1 object incurs a cost of c_1, then the total cost is equal to $\pi_0 c_0 (1 - Se) + \pi_1 c_1 (1 - Sp)$. The threshold on the ROC curve can be found which minimises this cost. It is not difficult to show that at such an optimal point the slope of the ROC curve must be $s = c_1 \pi_1 / c_0 \pi_0$. We can find the best point on the curve graphically by considering the line through the top left hand corner of the square and with slope $-1/s$. If we project the ROC curve onto this line, then the point which yields the projection nearest to the top left corner of the square is the point corresponding to the best threshold. This is illustrated in Figure 8.2. This idea can be used more generally to identify a range of threshold values which would be regarded as 'good'.

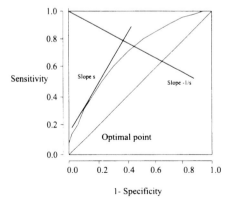

Figure 8.2 Graphical way of finding the optimal performance of a given classifier for specified costs.

ROC curves are all very well but, as we have remarked above, it would be nice if the continuum of comparative performances represented by such curves could be reduced to single numbers on which to base comparisons. Choosing a particular threshold is one way to reduce such a curve to a single criterion, but such an approach depends on the choice of the threshold (based, for example, on the assumption that one knows the relative costs of the different types of misclassification). An alternative criterion, which does consider the curve in its entirety, is given by the *area under the ROC curve*. If one curve dominated another then the dominant curve would have a larger such area. Of course, any reduction of two measures to a single measure is inevitably risky: if no curve dominates—if one is superior in some regions and another elsewhere—the area under the curve criterion will not help us to identify which is superior in any particular region.

The area under the curve is, in fact, equivalent to the two independent sample Wilcoxon–Mann–Whitney nonparametric test statistic; it gives the probability that a randomly selected class 0 subject will be identified as more likely to be class 0 than a randomly selected class 1 subject. As such, it appears to be a convenient and popular measure of separation between two distributions. It is also equivalent to the *Gini coefficient*, which is (usually) defined as twice the area between the curve and the diagonal. This is related to, but is not the same as, the Gini index mentioned in Chapters 4 and 6. The Gini index is a measure of how different the members of a set of probabilities are from each other, so it can be used as a measure of separability or impurity between distributions (as in Chapter 4) and hence as a classification rule performance criterion once a threshold has been selected. The Gini coefficient, on the other hand, provides a global summary of performance over all possible threshold values.

In practice, all of this has to be qualified by two things: (i) the fact that curves obtained from real data are not smooth curves but are step functions (except where there are ties, in which case the curves include diagonal line segments) and (ii) sampling variability considerations.

The first problem is typically overcome by plotting the curve by using straight line segments to connect points at intervals on the threshold scale so that several objects change classes between each threshold level. This is how Figure 8.1 was obtained. An alternative is to smooth the curve by fitting some parametric form. A popular, and predictable choice arises by assuming that the distributions of classifier scores for the cases and noncases are each normal. Other forms which have been used include logistic and negative exponential. Obviously there is a close relationship between such parametrised ROC curves and the curve of $f(1|\mathbf{x})$.

Sampling variability of ROC plots can be treated by plotting confidence bands instead of single curves. Statistical tests for comparing two ROC curves will depend on the design of the study. In particular, it will depend on whether the two curves were obtained by analysing two independent samples or by applying each classification rule to the same data. The latter is usually more statistically powerful.

A common approach to reducing the situation to a single degree of freedom is to attempt to define a cost function which accords appropriate relative weights to the two kinds of misclassification. As we have noted, a default choice is often to assume that these two kinds are equally serious so that the final measure becomes simply $(b + c)/n$, or error rate. Implicit in this is an assumption that the two types of correct classification incur no cost, an assumption which will generally be false. Any classification, whether it turns out to be wrong or right, will incur some cost. However, it is often reasonable to assume that this basic cost is the same whichever of the two classes the object belongs to and whichever class it is predicted to belong to (and then misclassificaton costs are added to this), so that it represents a constant added to every cell of the table. Being constant, it can be ignored in the analysis. (Section 1.2 mentioned the possibility that different subgroups of classes may be associated with different classification costs, but this does not apply in the two-class case.)

This default is far from ideal. It still represents a choice of the relative importance of the two types of misclassification, and hiding from the choice does not make it right! In some problems it may not be unreasonable to assume equal costs but, perhaps more often, it is risky. The ideal thing to do is to consult with experts in the domain of the classification problem to try to arrive at realistic relative costs. Techniques for extracting utilities may be applied here.

We use an example to illustrate the ambiguities which can arise if costs are not specified. Osteoporosis is a disease in which the bones deteriorate and crumble. It is a particular concern in postmenopausal women and is a major public health problem through its association with fractures of the hip, vertebrae and distal radius. It is thought that the most common manifestation in women under 75 is vertebral fracture. However, such fractures are often asymptomatic and diagnosis is often delayed until multiple fractures, height loss or spinal deformity have occurred. Treatments which may ease the condition do exist, but they are more likely to be effective the earlier they are started. At present the only reliable way of detecting vertebral fractures is by the invasive and relatively expensive technique of radiography. However, since several risk factors have been identified, a study was undertaken to see if an effective screening instrument could be constructed using a questionnaire based on these risk factors.

A sample of 1012 women aged between 48 and 81 were X-rayed and administered a questionnaire of some 45 items, exploring medical history, diet, exercise pattern and other risk factors. Various classification techniques were tried, including classical discriminant analysis, logistic regression, nearest neighbour methods, tree-structured methods and neural network methods. For our purposes it will suffice to examine just two of them, a logistic regression approach and a tree classifier.

Complete data were available for 851 of the women and we restrict ourselves to this subset because we are concerned only with comparisons between rules, and not the substantive problem. (If we were interested in developing a rule for real application, we would have to examine the missing values for patterns.) Of these 851, 65 (7.6%) were categorised as 'cases' and the remaining 786 (92.4%) as 'noncases' by a consultant radiologist's examination of the X-rays.

Table 8.2   Error rates (in parenthesis) for osteoporosis data.

		Logistic true class		Tree true class	
		Case	Noncase	Case	Noncase
Predicted class	Case	8	10	23	13
	Noncase	57	776	42	773
		(0.079)		(0.065)	

Table 8.2 shows the results of the two classification rules. The figures in parenthesis give the error rates, and we can see that the tree method is slightly better than the logistic method, although the statistical significance of the difference would need to be tested. However, as we have already stressed, such a comparison is based on an assumption that the two types of misclassification are equally serious. We have said nothing to justify such an assumption, and it seems unlikely to be the case. With different choices of relative costs, the relative performance of the two rules might be expected to differ. Indeed, for the sake of illustration, if we take the cost of misclassifying a noncase as a case to be 9 times as serious as the converse, the relative costs of the two classification rules are 14.7 for the logistic method and 15.9 for the tree method. That is, with this choice of costs, the tree method is worse.

Yet another way of reducing the situation to a single degree of freedom is by computing the ratio of sensitivity and specificity. In fact we shall take the ratio $Se/(1 - Sp)$. This is equal to $a(b + d)/b(a + c) = a/(a + c) \div b/(b + d)$, the likelihood ratio for the two classes. We see that, in terms of an ROC square, likelihood ratios are *slopes*. Such ratios tell us the relative risk of being classified into class 0 given that the subject actually comes from class 0 or class 1. Any point lying above the diagonal stretching from the lower left of the square will have $Se > 1 - Sp$, and hence will have a likelihood ratio greater than 1, meaning that cases from class 0 are more likely to be classified as class 0 than are cases from class 1. An ROC curve which passes below this diagonal gives cause for alarm.

A related single degree of freedom summary is the odds ratio (or cross-product), defined as the ratio of the odds of being classified into class 0 given that the subject actually comes from class 0 and class 1 respectively: $a/c \div b/d = ad/cb = Se.Sp/(1 - Se)(1 - Sp)$. Note that if class 0 has a very small prior (for example, if class 0 is a rare disease and the data are a random sample from the population) then $a/(a+b)$ and $c/(c+d)$ may be small so that $b/(a+b) \approx 1$ and $d/(c+d) \approx 1$. It follows that $ad/cb = (a/a+b)$ $\times (d/c+d)/(c/c+d)(b/a+b)$ $\approx (a/a+b)/(c/c+d) =$ (positive predicted value) / (1−negative predicted value).

Often the log transform of the odds ratio is more convenient to work with. One property which is worth noting is that s.e.[ln(odds ratio)] $\approx \sqrt{1/a+1/b+1/c+1/d}$, from which approximate confidence intervals and tests can be constructed.

Sometimes the context of the problem provides a natural way to reduce the two degrees of freedom to one. In particular, the two classes are often defined in terms of possession of a property and lack of that property—having a disease versus not having the disease, signal present or absent, a good insurance risk versus a poor risk, and so on—and other constraints mean that only one type of misclassification is of interest. For example, in some tax evasion investigations only a given fixed proportion of the population can be investigated, so the aim is to minimise the proportion misclassified into that class. In terms of Table 8.1, with a prespecified proportion being classified into class 0 (if this class represents the ones thought to be worth investigating), the criterion to be optimised is simply $a/(a+c)$. This has interesting and, perhaps at first, unexpected implications.

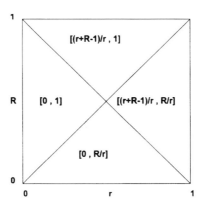

Figure 8.3 Bounds on proportion of goods amongst those accepted when proportion r are accepted and proportion R are good.

Suppose that a proportion r of the population has been designated as to be accepted for further investigation. Suppose also that a proportion R of the population are known, from previous studies, to be good risks. If $r < R$ it might be possible to construct a classifier which would have 100% good risks amongst those accepted. That is, 100% is the upper bound on what can be achieved. On the other hand, if $r > R$ the best that could possibly be achieved would be a good rate, amongst those accepted, of R/r. Similarly, if $r < 1 - R$ the worst that could ever be achieved would be 0% correct, whereas if $r > 1 - R$ the worst that could be achieved would be $(r + R - 1)/r$. These four bounds together partition the (r, R) square into four segments, as shown in Figure 8.3. This is quite different from the conventional problem where error rate is the criterion of interest. In principle, the a priori bounds on error rate are always 0 and 1, although fixing the proportion to be accepted will also impose bounds on the error rate (cf. the computations involved in calculating Fisher's exact test). In one problem which we examined (Henley, 1995; Henley and Hand, 1996), concerned with classifying applicants for credit as 'good risks' or 'bad risks', it was known that about 45% of the population in question were good risks and it was predetermined that 70% of the applicants would be accepted. Applying the above bounds, we see that the proportion of good risks amongst the accepted applicants must lie in the interval [0.22, 0.65]. A classifier which achieved a rate of 60% goods amongst those accepted in this problem might be thought of as not very good. The bounds, however, show that it is performing almost optimally given the constraints upon it. A similar calculation shows that, for this problem with these constraints, the error rate must lie between 0.25 and 0.85. Again, a 25% error rate might be thought of as poor; such a classifier misclassifies 1 in 4 of the applicants. However, this is the best that can possibly be done under the circumstances.

The fact that there are many ways of summarising the four numbers in the body of Table 8.1 tells us there are many ways of looking at performance, even if the situation is intrinsically two-dimensional. An example will illustrate this, and the care which has to be taken.

Table 8.3   Data on dance students and anorexia nervosa. (From Garner and Garfinkel, 1980)

EAT score	Anorexics	Controls
≥ 30	11	58
< 30	1	113

Table 8.3 shows data reported by Garner and Garfinkel (1980) concerning dance students, each of whom is known to be either anorexic or not anorexic (a 'control' subject), classified according to their score on the Eating Attitudes Test (EAT), an instrument used to screen for anorexia nervosa. A score of 30 or more is taken to be indicative of the illness. From the table we can estimate the sensitivity as $11/(11 + 1) = 91.7\%$ and the specificity as $113/(113 + 58) = 66.1\%$. Before we proceed further, this gives us an opportunity to remark that what are regarded as 'high' values vary from discipline to discipline. In screening for diseases like anorexia, figures such as those above are

respectable. In contrast, in certain hard science or engineering classification applications, figures below the high 90s would be regarded as poor. In the present example, from the figures above, one might conclude that the EAT is an effective classifier, certainly for screening purposes.

However, we can also look at the predicted values of the test. That is, the proportions which really do come from class i amongst those predicted to have come from class i. From Table 8.3 we see that the positive predicted value is $11/(11 + 58) = 15.9\%$. This means that of 100 students classified by the EAT as anorexic only 16 will in fact be so (subject to sampling variability, of course). A situation like this could be quite serious. At first glance, one might regard misclassifying ill people into the well class as the more serious type of misclassification (according it the higher cost) because that means that they will not be treated. But the converse can also be serious; it may mean that well people undergo unnecessary medical treatment or intervention, which carries its own financial costs and risks.

How can it happen that one pair of measures (which summarise the entire two-dimensional space of variation) shows the classifier to be good whereas some other measure shows it to be bad? The answer, of course, is that the measures are related via the other numbers appearing in Table 8.1. In the present example, we have that the positive predicted value (pp) is

$$pp = \pi_0.Se/[(1 - Sp) + \pi_0.(Se + Sp - 1)]$$

From this we see that low prevalence (the *prevalence* of a disease in a population is the prior probability of that disease in that population) can produce a low positive predicted value even if the sensitivity and specificity are large. For example if $\pi_0 = 0.01$ and $Se = Sp = 0.9$, then pp is only 0.09. Similarly, with $\pi_0 = 0.01$, to obtain $pp=0.9$, we need $Se=Sp=0.999$ (taking $Se=Sp$ for simplicity of exposition). In areas such as psychiatric screening, a typical relationship between the values of Se and Sp is that $Se+Sp\approx1.6$. Adopting this relationship, Table 8.4 shows the dramatic way in which pp will decrease with Sp when $\pi_0 = 0.01$.

There is another way of looking at the relationship between sensitivity and specificity on the one hand and predicted values on the other. Sensitivity and specificity tell us about the performance of the classifier as a whole—how good it

Table 8.4 Relationship between Se, Sp, and pp when $\pi_0 =0.01$.

Se	Sp	pp
0.6	1.0	1.00
0.7	0.9	0.15
0.8	0.8	0.04
0.9	0.7	0.03
1.0	0.6	0.02

is at detecting a case and how good it is at detecting a noncase. The predicted values tell us how good the classifier is at classifying an individual subject correctly. For example, subjects classified into class 0 have a probability given by the positive predicted value of really being class 0 (again, subject to sampling variation). Clearly extreme care is required when choosing a performance criterion to ensure it is the most relevant. This point is brought home in the following section.

8.2   FITTING CRITERIA TO THE OBJECTIVE

We noted, in Section 6.1, that there may be several fundamentally different reasons for wanting to build a classifier. These different reasons will, presumably, be reflected by different criteria for assessing performance. It then follows that the different performance criteria should be reflected by different criteria to be optimised in constructing the classifiers. In this section we illustrate how different reasons manifest themselves in different criteria. In particular, we compare classification when the objective is *diagnosis* with classification when the objective is *screening* and with classification when the objective is *prevalence estimation*. Although we have used medical terms to distinguish between the three situations, they are quite general and crop up in all sorts of contexts. Similarly, although we only distinguish three different objectives, this does not mean that there are no others. Sufficient care to distinguish between these objectives is not always taken. This means that sometimes classification rules are constructed with one objective when the aim is really another. The potential consequences in terms of the accuracy and validity of the results are obvious. We begin by defining our terms.

- *Diagnosis* aims to assign an individual to the class to which it has the greatest probability of belonging.
- *Screening* aims to identify that part of the population most likely to belong to each of the classes.
- *Prevalence estimation* aims to estimate the prior probabilities of the classes— their 'sizes' in the population.

Let us first contrast diagnosis with screening. As a premiss we assume that we know the class priors, at least approximately.

In diagnosis the centre of interest is the individual; the aim is to find the class to which the individual in question has the greatest probability of belonging. To do this we need to estimate the probability that the individual in question comes from each class. This is simply $f(j|\mathbf{x})$ for the jth class. We choose the largest of these and classify the individual into the corresponding class. How other individuals may be classified is irrelevant to the classification of this individual. Especially in classes with small priors, it can happen that, for some j, $f(j|\mathbf{x})$ is never the largest, for any $\mathbf{x}$. This would mean that no individuals would ever be classified into class j; they are more likely to have come from some other class. This is perfectly sensible from the perspective of the individual. In the two-class case, denoting $f(1|\mathbf{x}_i)$ by p_i, the likelihood is $\prod_i p_i^{\alpha_i}(1 - p_i)^{1-\alpha_i}$ where $\alpha_i = 0,1$. This is maximised by putting $\alpha_i = 1$ if $p_i > 1/2$ and $\alpha_i = 0$ otherwise. If $p_i > 1/2$ for all i, no individuals would be assigned to class 0.

In contrast, in screening, the focus of our interest is not the individual, but is the population as a whole. We are trying to identify that *part* of the population which is most likely to belong to each class. We assumed above that we knew the class priors. This means that we know *how many* subjects we need to classify into each class; the problem is merely to identify which particular ones go where. The diagnostic solution above is generally unacceptable: in classifying each individual we must take into account how many of the others have been classified into each class. Now, the mere fact that an *individual* is most likely to have come from class j does not mean they will be classified into that class. In the two-class case the likelihood is the same as above but subject to the constraint that $\sum_i \alpha_i = \pi_1 n$, where n is the total size of the population to be screened. In the two-class case we need to rank the probabilities and assign those $\pi_1 n$ with the largest probabilities of belonging to class 1 to class 1, regardless of whether or not these probabilities are greater than 1/2.

In the osteoporosis example, we showed that ambiguity about cost functions led to ambiguity about which of two classifiers was superior. In fact, the objective of developing a classification rule in that example was really screening. We should have imposed a constraint on the classification: that the number to be classified into each class is equal to the numbers known to have come from that class. This would have given us the appropriate threshold for classifying subjects in future. And, yet again, we emphasise that all this is subject to considerations of sampling variability.

Table 8.5 Screening classifiers for the osteoporosis data.

		Logistic true class		Tree true class	
		Case	Non-case	Case	Non-case
	Case	16	49	37	28
Predicted	Noncase	49	737	28	758
class		(0.115)		(0.066)	

Table 8.5 shows the confusion matrices which arise for each of the classifiers when we do this. They differ from those shown in Table 8.2. In each case, the correct number, 65, are classified into the 'case' class. Now the tree classifier appears to be substantially better than the logistic classifier.

Table 8.5 also shows us another property which arises in the two-class case when the aim is screening. From the above, we will choose the threshold for the classification rule so that $a + b = \pi_0$. By definition, $a + c = \pi_0$, so we have that $b = c$. This is apparent in both subtables of Table 8.5. It immediately follows that the overall relative costs of the two subtables are invariant to the costs assigned to the different types of misclassification: if misclassifying a class 0 individual incurs a cost of r and misclassifying a class 1 individual incurs a cost of s, the overall cost is $rc + sb = (r + s)b$ when $b = c$. That is, the relative cost is simply proportional to the number of individuals misclassified (or, equivalently, the number in either of the two misclassification cells of the table).

The rules which result from each classification method when we are interested in screening are minimax rules; they are such that the maximum number of misclassifications of each type is minimised.

In fact, this point generalises. Without loss of generality put $s = 1 - r$, let one classifier have confusion matrix as given in Table 8.1, with entries a, b, c and d, and let another have entries a', b', c' and d'. Then the difference in costs between the two classifiers is $r(c - c') + (1 - r)(b - b')$. Now suppose that we want to classify a certain number as 'cases', without restricting this number to be equal to $a + c$. Then $a + b = a' + b'$ and we already know $a + c = a' + c'$. It follows that $c - c' = b - b'$. Hence the difference in costs between the two classifiers is simply $b - b'$, regardless of the value chosen for r. We should be clear about the nature of this invariance. For a given number to be classified into each class, the cost difference between the two classifiers is not dependent on the misclassification costs (given that $s = 1 - r$). However, if the number to be classified into each class is changed then the cost difference will change (and could invert).

So much for diagnosis and screening. Now let us include prevalence estimation in the equation. One important aim of epidemiological work is to estimate the disease prevalence (prior) within a population. Typically it is impracticable to examine the entire population, so a sample is taken. Often it is impossible to determine the true class of each person in the sample (perhaps it would cost too much, or perhaps determining the true class requires a post-mortem examination, and so on) so a subsample is taken and the true class determined for this smaller subsample. A simple but less accurate classification rule is then applied to the complete sample, and its performance is calibrated by comparing its performance with the true class on the subsample. The important thing to note here is that, in applying the classifier to the sample, the aim is not to classify each sample member as accurately as possible (i.e. the aim is not diagnosis). Nor is it to identify which part of the sample is most likely to have come from each class (i.e. the aim is not screening). Instead the aim is to find the most accurate estimate of a population parameter, the prevalence.

To illustrate, contrast a classifier aimed at minimising error rate with one aimed at estimating prevalence with the smallest variance in the estimate. Error rate can be written as $\hat{e} = fn.\pi_0 + fp.\pi_1$, where fn and fp are the false negative and false positive rates respectively. The prevalence estimate is

$\hat{p} = (1 - np)p_1 + pp.p_0$, where np and pp are, respectively, the negative and positive predicted values. The predicted values will be obtained from the cross-classified subsample, and the p_j will be obtained from the larger sample, classified solely by the 'inaccurate' classifier (and we assume that variation in the p_j contributes negligibly to the overall variation). The variance of this prevalence estimate can be written in several ways, including $\mathrm{var}\,(\hat{p}) \propto fp.Se + fn.Sp$. This is quite different from $\hat{e}$, so it is hardly surprising that classification rules built for the two purposes may differ.

As an example, suppose that the cases and noncases are normally distributed, $N(0,1)$ for class 1 (the noncases) and $N(2,1)$ for class 0, with the prevalence (prior of class 0) being 0.1. Then the minimum variance of the prevalence estimate is proportional to 0.267. This increases to 0.538 at the classification threshold which minimises the error rate. Conversely, the minimum error rate is proportional to 0.070, and this increases to 0.159 at the classification threshold which minimises the variance of the prevalence estimate. The optimal classification threshold for one use is not optimal for the other. In general, a rule built for one purpose may not be very good for some other purpose, so the objective must be determined before the rule is constructed.

8.3   FROM SUMMARY MEASURES TO OVERALL COMPARISONS

An ROC curve shows what the performance would be like at each possible threshold, but to produce a 2×2 confusion matrix such as that in Table 8.1 it is necessary to settle on a particular threshold. Once this has been done, the classification rule is complete and its performance can be assessed and compared with other rules via measures such as sensitivity and specificity. Sometimes, however, we want to go in the reverse direction: we are given summary measures arising from a particular (unknown) threshold and we would like to produce performance measures at other thresholds. For example, we might be given the sensitivity and specificity of a rule and we might want to produce an entire ROC curve. Or we might want to summarise the performance of a rule as estimated in several different studies, possibly using different (and unknown) thresholds. To do this we would need somehow to transform the results to a common threshold. Such an objective arises in meta-analytic studies.

Clearly it is impossible to go from two summary statistics to an entire curve unless constraints are imposed on the possible shape of that curve; equivalently, it is impossible unless assumptions are made about the forms of the distributions involved. First, let us assume that the two populations have logistic distributions with equal variances. Let the dispersion parameter of the two classes be β (so the variance is $\pi^2\beta^2/3$) and, for convenience (it will not detract from the generality of the results), we will assume that class 0 has mean 0 and class 1 has mean μ. Then the distribution function for class 0 is $(1 + e^{-x/\beta})^{-1}$ and for class 1 it is $(1 + e^{-(x-\mu)/\beta})^{-1}$. From this it follows that a, b, c and d are respectively proportional to $(1 + e^{-k/\beta})^{-1}$, $(1 + e^{-(k-\mu)/\beta})^{-1}$, $1-a$, $1-b$, where k is the unknown threshold. Hence the log odds ratio is $ln(odds) = ln[Se.Sp/(1 - Se)(1 - Sp)] = \mu/\beta$. In terms of the standard deviation of the logistic distributions, this is μ/σ. $\pi/\sqrt{3}$. That is, if we assume logistic distributions, the log of the odds ratio is proportional to the standardised difference between the means of the two distributions. The two measures, sensitivity and specificity, obtained from an unspecified threshold, have been reduced to a single measure which is independent of that threshold and which can be used to compare classification rules.

More generally, assume that the corresponding cumulative distributions are $V_i(x)$ with respective means μ_i. Then, with classification threshold k, the sensitivity is $Se = V_0(k)$ and the specificity is $Sp = 1 - V_1(k)$. Now define distributions $W_i(x)$ to have the same shape as the V_i distributions but with means 0. That is $W_i(x) = V_i(\mu_i + x)$. Then it is easy to show that $\mu_0 - \mu_1 = W_1^{-1}(1 - Sp) - W_0^{-1}(Se)$. Again this is invariant to the choice of threshold.

8.4 FURTHER READING

A comprehensive description of ROC analysis is given in Zweig and Campbell (1993). Hilgers (1991) describes confidence bounds for ROC curves. Hanley and McNeil (1982) give standard errors and tests for areas under ROC curves. Metz *et al.* (1984), for parametrically smoothed approaches, and Delong *et al.* (1988) for raw data based approaches, describe tests to compare the areas under two ROC curves when they have been obtained from the same subjects. Wieand *et al.* (1989) also consider this situation and describe a range of test statistics, obtained by integrating the difference between the sensitivities of the classifiers over a range of values of the specificity, with various weightings within that range. Their methods also apply to the case when the two classifiers are evaluated on independent groups.

Henley (1995) and Henley and Hand (1996) describe bounds on performance rates arising from optimising criteria which do not include both types of misclassification (class 0 into class 1 and vice versa). Williams *et al.* (1982) describe the problems of screening for uncommon disorders in more detail, with special reference to the Eating Attitudes Test. The distinction between screening and prevalence estimation is discussed further in Hand (1987b, 1995b) and Pickles *et al.*(1995), illustrated by data from the General Health Questionnaire. The osteoporosis study outlined above is described in more detail in Cooper *et al.* (1991), Hand (1994a) and Hand (1995a).

Hasselblad and Hedges (1995) and Hand (1995a) describe methods for transforming classification rule summary statistics, based on unknown thresholds, to single commensurate measures of separability between the distributions involved. Hand (1995a) presents general results for arbitrary assumed distributional forms. Hasselblad and Hedges (1995) are particularly concerned with the use of such summaries in meta-analysis and they also explore the consequences of departures from an assumption of logistic distributions with equal variance.

Part IV

Practical Issues

CHAPTER 9
Some Special Problems

9.1 VARIABLE SELECTION

This chapter describes some further classification problems which, though perhaps important, did not justify entire chapters of their own in a book of this length; we begin with variable selection.

In principle, one could use all of the variables one has measured. This can mean that the measurement space has a very high dimensionality, especially in certain classes of problems. Because of this one might like to reduce the dimensionality simply in order to make the classification calculations quicker, to save storage space or to permit rapid communication of the measurement vectors. However, a more fundamental reason for choosing a subset of variables arises from the curse of dimensionality and overfitting issues discussed in Chapter 1. With a very large number of variables, the design set is unlikely to generalise well. A plot of performance against dimensionality generally shows improving performance in terms of the design set (for example, apparent error rate or quality of fit of model to the design data) but initially improving and then deteriorating performance in terms of a test set (this is sometimes called *peaking*; see Hand (1981a, Chapter 6) for illustrations. As a result, one will be interested in reducing the size of a large-dimensional problem.

There are two basic strategies one might adopt for identifying a low-dimensional subspace of the original measurement space. The first is simply to select a subset from the original set of variables. The second is to define (a few) new variables as (transformations of) combinations of the original ones. The first has the merit that, once one has identified the subset in question, only the variables in this subset need to be measured on future objects. The second has the merit that it is more flexible, so it may lead to more powerful classifiers. Of course, as we have noted elsewhere, the separation implied by the second method, into feature extraction and classifier construction, is really artificial. They are really both aspects of the overall process of constructing a model for the decision surface (or probabilities).

As far as selection goes, perhaps the most obvious approach to choosing a subset of variables is to examine the variables separately and to choose the few best (in terms of impurity or separability). However, the best k individual variables need not be the best set of k. That is, the subset of k produced by this procedure may not be the best subset of k which could be produced. This is easily seen by considering the extreme case in which the k variables are effectively identical. Then taking all k of them would add very little information beyond that due to the first; one would do better by taking some other, less highly correlated variable, even though, by itself, it may have less predictive power in terms of class separation.

Moving beyond this naive approach, one might consider examining all possible subsets of variables from the entire available set (or, perhaps, all possible subsets of a given size). However, the number of potential variable sets which could be chosen from a given total number of variables can be vast. In principle, to find the global best, each of these sets would need to be examined and its impurity estimated. This means that measures of impurity which can be calculated rapidly (such as those based on only first-and second-order functions of the sample data) are attractive. It also means that ways to accelerate the search have been developed. They can be divided into two types: those which restrict the search to a subspace of the entire space of possible variable sets, and those which use sophisticated methods to accelerate the search.

The first type includes classical stepwise methods. The basic approach here begins by comparing all the variables individually and choosing the best, where 'best' is determined by some measure of impurity or separability. Then all the remaining variables are examined to identify that one which, when combined with the first, yields greatest between-class separability. This is repeated, at each step adding that variable which, when combined with those already chosen, leads to the best results. Note that a set of k variables chosen in this manner may not be the best set of k. Many potential subsets will not have been examined at all by this procedure.

This basic method has been improved in several ways (at the expense of increased computation time). For example, at each stage one might consider adding not the individual variable which leads to greatest improvement, but the pair of variables which do. Increased computer power has also led to parallel methods; at each stage one retains not the single best set so far found (by the stepwise procedure) but the best few sets. Each of these is then expanded (by whatever method is being used) and the overall best few retained for the next stage.

Backwards methods are also used: one starts with all the variables and sequentially removes that one whose removal leads to least degradation in performance. Backwards methods are more computationally demanding, since the impurity calculations relate to higher-dimensional subspaces. Backwards methods have also been extended to permit more than one variable to be deleted at each step. Of course, one can combine forward and backward stepwise methods. For example, add those two variables which lead to greatest improvement and delete that one whose removal leads to least degradation.

None of these stepwise methods guarantees finding the globally best subset of variables. However, methods other than exhaustive search which guarantee this do exist, provided the original number of variables is not too large and provided the measure of impurity possesses certain properties. For example, *branch and bound* methods allow one to eliminate entire sets of sets of variables, without explicitly considering them. Suppose, for example, that the impurity measure decreases monotonically with increasing size of variable set (so that adding a variable to some set necessarily decreases, improves, the impurity measure). Suppose that we want a variable set of size 2 and have determined that the set $\{a, b\}$ has impurity measure 0.2. Suppose we are now considering a set $\{c, d, e\}$, and that this has measure 0.7. Then there is no point in looking at two-variable subsets of $\{c, d, e\}$. Each of these must necessarily have impurity values no less than 0.7, which is worse than the set we have already found. Branch and bound methods are described in Narendra and Fukunaga (1977), Hand (1981a,b) and Ridout (1988).

Since variable selection is a search or optimisation problem, many other approaches have been explored, including various methods of mathematical programming.

Feature *extraction*—defining new variables on which to base the classification rule by transforming (combinations of) the raw variables—received a great deal of attention in work on statistical pattern recognition in the 1960s and 1970s, where it was essentially treated as a preprocessing stage prior to the construction of the classification rule or decision surface. But this is a rather artificial distinction, with the artificiality especially brought home in the contexts of (a) classical linear discriminant analysis in the two-class case, and (b) neural networks. Linear discriminant analysis extracts a single linear function, requiring only a threshold to be added to define the decision surface, so it is difficult to say where 'feature extraction' ends and 'decision surface construction' begins. Neural networks integrate the two processes into one, with internal nodes of the network effectively defining features. Devijver and Kittler (1982) contains a good account of the pattern recognition work on feature extraction.

The key to effective feature extraction is finding a good measure of separability or impurity. Here 'good' means that, as well as being related to the criterion in which one is really interested, the chosen measure should be easy to calculate and feasible to optimise. Measures differ according to the relative weights they assign to these desiderata. They range in complexity from simple measures, which can be optimised analytically (such as the ratio used in classical linear discriminant analysis), to complex measures requiring integration, as discussed in Chapter 6. Linear functions of the raw measurement variables, in particular, have received much attention, because of the ease of calculation. For example, a common approach is to take the first few principal components of the space spanned by the predictor variables, where they may be ordered according to their variance or according to their correlation with the criterion (class) variable. Some such approaches are briefly discussed in Section 2.6.

Finally, as a cautionary comment, when there are a large number of variables from which to choose and few design set objects to help in that choice, there is clearly a great potential for overfitting in the selection or extraction process. This can be alleviated by cross-validation.

9.2 HIGH-DIMENSIONAL ALLOCATION RULES

It is useful to distinguish two different kinds of high-dimensional allocation problems. In one (often arising with automatic data collection via electronic measuring instruments) the variables correspond to measurements of the same thing at different places on an underlying continuum (such as time or wavelength). As a consequence, successive variables are often highly correlated. For example, near-infrared spectroscopy may measure absorbances at 1000 different wavelengths. In such cases although many variables have been measured, the correlations mean that the information summarising the shapes of what are effectively profiles can be described in terms of a much smaller number of *features*. But it may be far from trivial to find these features—to extract the information which describes how the profiles belonging to different groups differ. Methods which do this automatically, such as neural networks, may be particularly useful in this situation. Since the raw variables are highly correlated, it may be possible to devise effective forms for the covariance matrix with far fewer degrees of freedom than are needed for a completely unspecified matrix. Section 2.5 describes methods based on this idea. Problems in which the variables are highly correlated may be viewed as being relevant in only restricted regions (and subspaces) of the overall measurement space. This means that, although such problems may initially appear intractable due to the high dimensionality, this may not be the case in practice.

The other class of problem occurs when there is no underlying continuum inducing a high degree of correlation between successive variables. For example, management consultants may use very large questionnaires to explore feelings, opinions and behaviour within an organisation. In cases like this, no doubt there will be high correlations between variables, but the expected pattern for such correlations may not be so straightforward to determine. Again, Section 2.5 outlines some basic ideas for this sort of problem.

The mere fact of having a large number of variables in itself causes no problems for the construction of classification rules in principle; in practice there may be difficulties over to having to invert very large matrices or due to the time it takes to estimate parameters. However, when the number of dimensions is large, especially if the correlations between them are weak, a very large number of design set objects could be needed to provide adequate coverage of the measurement space. Without such large sets there are severe dangers of overfitting (resulting in poor generalisation), as discussed in Section 9.1. Often, however, few design set objects are available. For example, Stone and Jonathan (1994), in discussing 'quantitative structural–activity relationship studies' (QSARs), say: '(i) Most QSAR studies are based on a fixed number of compounds, n, which is rarely in excess of 100 and usually less than 50. (ii) There is often effectively no limit to the value of the number of variables, p.' In such situations one standard class of ways to tackle the problem is to reduce the number of variables using the methods described in Section 9.1.

Another way to alleviate overfitting problems is to adopt a fairly inflexible model type (of the kind one hopes will provide an adequate representation of the underlying distributions). For example, a common approach is to base things on only first-and second-order statistics of the measurement vectors. Various techniques along these lines have been developed. They include partial least squares methods (finding that linear predictor which has maximum covariance with the class variable), shortest least squares (which imposes an extra constraint on the solution using a generalised inverse in order to produce a unique solution), ridge regression (which biases the estimate away from the ordinary least squares estimate), SIMCA (which models the similarity of the point to be classified to each of the classes) and DASCO (a refinement of SIMCA). Since these are all based on covariance matrices of the predictor variables, we have grouped their descriptions together in Section 2.6.

Finally, one may try to overcome the problem by fitting a flexible model and regularising it—effectively smoothing away its more dramatic irregularities.

9.3   THE FLAT MAXIMUM EFFECT AND SIMPLE WEIGHTED SUMS

Linear classifiers have several attributes which make them very popular: they are simple to construct and use, they are easy to explain, they produce classifications quickly and they are often surprisingly accurate. They produce a classification by comparing a weighted sum (a linear combination) of the measurements with a threshold. The weights in the sum represent how important each variable is in determining the overall score, when taken in conjunction with the other variables in the sum; differences in a variable with larger weights will have a greater impact on the overall score than will the same differences in a variable with small weights. In fact this does not necessarily imply that those with larger weights will have a greater impact on the *classification* performance. This will depend on the position of the decision surface and the distributions of the objects.

A special case of the linear classifier arises in *point scoring* rules. In these, the measurements are (typically) binary, taking values 0 or 1, and the final score is produced simply as the sum of the measurements, equivalent to the number of them which take the value 1. Examples are given in Burgess (1928) and Greenwood (1982). The fact that the variables are simply summed means that, in terms of the above linear form, the weights are all taken to be equal. This might lead one to suppose that a substantially better (albeit rather more complicated) linear rule could be developed using the optimal (and generally not equal) weights. However, it has been observed in applications that the predictive power of linear rules is often insensitive to the precise values of the coefficients. This property has been described as the *curse of insensitivity* (Rapoport, 1975) or the *flat maximum effect* (von Winterfeldt and Edwards, 1982). The negative implications of the first term arise from situations where the aim is to *understand* rather than merely to predict. Then, with the aim being to place an interpretation on the weights in the linear combination, accuracy of the estimates of the weights and uniqueness of their definitions is important. In this book, however, the aim is *prediction*, not interpretation, and such insensitivity can be more of an advantage than a disadvantage. (Another way of putting this distinction is that the weights in the models discussed in this book are *used* whereas those in the other class of models are merely *interpreted* (Green, 1977)).

Wainer (1976) proved the following. Suppose that the p predictors are scaled to have zero mean and unit variance, that the response is similarly scaled, that the predictors are independent and that the coefficients in a multiple regression are uniformly distributed between 0.25 and 0.75. Then if all the regression coefficients are replaced by the value 0.5 (i.e. all equal), the expected loss of variance will be less than $p/48$. In fact, Wainer had $p/96$; see Laughlin (1978) for the correction. Pruzek and Frederick (1978), however, argue that Wainer's conditions are in fact highly restrictive. Since the variables are standardised, the regression coefficients equal the corresponding correlation coefficients. Moreover, since the predictors are independent, the sum of squares of the correlation coefficients is equal to the squared multiple correlation coefficient, which is bounded above by 1. It follows that the average value of the squared individual correlation coefficients is less than $1/p$. With all the coefficients in the range [0.25, 0.75], the maximum value that p can take is then $1/0.25^2 = 16$, occurring when all coefficients equal 0.25. If the average value of the coefficients is 0.5 then p is bounded above by 4.

Wainer (1976) extended his argument to the case when the predictors are not independent, but Pruzek and Frederick (1978) argued that the specific relationships between the regression coefficients must be taken into account 'before one can know how well equal weights will work in relation to idealised weights.' (They then go on to develop a system which takes account of one's prior beliefs about the regression coefficients.) Despite all this, it does seem that the equal weights approach can be suprisingly accurate in many practical applications (see also Dawes and Corrigan, 1974; Wainer, 1978). At least part of the explanation may lie in overfitting of the regression weights to the design sample; the objective, after all, is to make predictions for *future* cases. In any case, practical issues other than crude classification accuracy need to be taken into account. (i) The relative merits of slightly greater classification accuracy must be balanced against greater complication; one of the attractions of linear rules is their simplicity. (ii) Populations evolve over time and in most applications the data to which a rule will be applied are collected after the training set data. (iii) Overstreet and Bradley (1995) gives an example of a generic rule being applicable to a range of problems, which clearly would not be the case if the rule were very finely tuned to one of the problems. (iv) In Section 9.4 we compare classes defined by partitioning a continuum with those defined according to a real qualitative distinction. In the former case a slight change in the defining continuum would mean a shift in the decision surface. Accuracy according to one definition represents slight inaccuracy according to another, and yet each definition is equally legitimate.

9.4 CLASSES FORMED BY PARTITIONING A CONTINUUM

Classical linear discriminant analysis with two classes, as described in Chapter 2, is optimal if the distributions of the measurements are ellipsoidal for each of the two classes and provided they have equal covariance matrices. The most important special case of this arises with the multivariate normal distribution. This assumption is reasonable if the two classes are qualitatively distinct, such as males versus females or two different letters of the alphabet in character recognition. But sometimes the classes are not qualitatively distinct; they are defined by imposition of a cutoff point on some underlying continuum. Examples of this are old versus young, where the underlying continuum is age, and the common medical classification into mild or severe, where the underlying continuum is severity of illness. Sometimes the score on the latent continuum is never explicitly produced. Indeed, it might be that explicitly measuring this score is very difficult or even impossible, so that only the classification is observed (this typically occurs in the medical example and in questionnaires which require the respondent to answer a question by ticking one of a ranked set of boxes).

In those situations where an underlying continuum exists, a reasonable assumption would be that the joint distribution of the measurements and this underlying continuum is multivariate normal. If this assumption is made, it follows that the marginal multivariate distribution of the measurements for each of the classes separately *cannot* in general be ellipsoidal and cannot, for example, be multivariate normal. Since such a situation arises often, a pertinent question is, How would linear discriminant analysis be expected to perform in these circumstances? In particular one may ask, How does the decision surface produced by this common method compare with the optimal decision surface? These questions are answered in Hand *et al.*(1996), summarised below.

The decision surface produced by classical linear discriminant analysis (Section 2.2) is

$$\mathbf{x}^T \Sigma^{-1} (\mu_1 - \mu_0) = \ln p_0/p_1 + (\mu_1 + \mu_0)^T \Sigma^{-1} (\mu_1 - \mu_0)/2$$

where μ_j is the mean of class j, Σ is the assumed common covariance matrix of the classes, and p_j is the prior for class j.

In contrast, when the two classes have been defined by partitioning a multivariate normal distribution, the optimal decision surface is given by the line of intersection of the regression plane and the plane orthogonal to the response axis at the threshold. That is, the optimal decision surface is

$$\mathbf{x}^T \Sigma_{xx}^{-1} \Sigma_{xy} = (t - \mu_y) + \mu_x^T \Sigma_{xx}^{-1} \Sigma_{xy}$$

where the terms are given by the parameters in the joint multivariate normal distribution

$$f(y, \mathbf{x}) = N\left(\left\{ \begin{matrix} \mu_y \\ \mu_x \end{matrix} \right\}, \left\{ \begin{matrix} \Sigma_{yy} & \Sigma_{yx} \\ \Sigma_{xy} & \Sigma_{xx} \end{matrix} \right\} \right)$$

with y the response and $\mathbf{x}$ the vector of predictor variables.

The question we want to answer is, How does the first of these rules, when the classes are formed by dividing the y variable at some threshold, compare with the second, optimal, rule?

There are two aspects to the comparison: the orientation of the decision surfaces and their position. Hand *et al.* (1996) show, firstly, that the decision surfaces are parallel and, secondly, that the constants (the classification thresholds) for the optimal and linear discriminant analysis solutions are respectively given by $c_o(t) = -t$ and

$$c_{LDA}(t) = \sqrt{2\pi}\, e^{-t^2/2}\, p_1 p_0\, \ln\left(\frac{p_0}{p_1}\right) + \frac{R^2 e^{-t^2/2}}{\sqrt{2\pi}} \left[\frac{1}{2}\left(\frac{1}{p_1} - \frac{1}{p_0}\right) - \ln\left(\frac{p_0}{p_1}\right)\right]$$

where R^2 is the squared multiple correlation coefficient between the $\mathbf{x}$ variables and the underlying continuum y, and where t is the threshold at which y is cut.

An illustration of the practical implications of this result is given in Figure 9.1, for the case of $R^2 = 0.95$. The upper panel shows plots of error rate against t and the lower panel the ratio of the two error rates against t. The optimal rule produces an error rate of only about 55% of that for the linear discriminant analysis classification rule when t is about 2 (so the optimal error rate is about 1%). For smaller values of R^2 the difference between the two rules is less marked.

9.5   REJECT INFERENCE

In certain situations the design sample is not simply a random sample from the mixture distribution of the classes, but is distorted in some way. This arises, for example, in medical problems involving a rare disease, where sampling according to the true class priors would yield very few cases, so that effective classification rules could not be constructed. This is overcome by oversampling from the rare class, then adjusting the derived classification rule accordingly. It also happens in screening problems when a new classifier is to be developed. Apart from any original training data which may be available (which was used for

Error rates

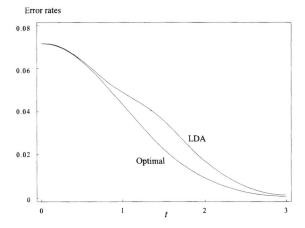

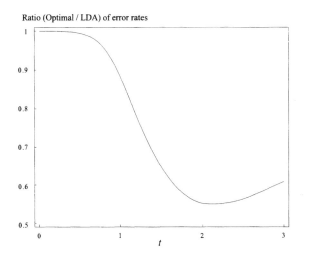

Figure 9.1 Classification when $R^2 = 0.95$: (Top) error rates for the optimal and linear discriminant (LDA) classification rules, (Bottom) the ratio of the two error rates, optimal/LDA.

constructing the original instrument), new data will only have true classes known for those subjects classified as cases by the original instrument. For other subjects, their measurement vectors will be known, but their true classes will be unknown. This situation also arises in credit scoring, where the aim is to classify an applicant for credit as a good or bad risk (see Section 10.3). In this situation the original training data is typically unavailable, so the new rule must be constructed solely from the distorted sample: a sample in which the true class is unknown for those regions of the measurement space which the original classifier decided were bad risk regions. The term *reject inference* is used to describe inferences about the true class of applicants in such regions, since such applicants have been rejected rather than accepted for a loan. One important reason for attempting such inferences is the possibility that they may be useful in constructing an improved classifier.

We can distinguish two distinct situations. In the first there are regions of the measurement space containing no design set points. This arises, for example, if the set of predictor variables includes all those variables originally used in the selection of the design sample. Then the original variables span a marginal subspace of the space of the new predictor variables. The second situation arises when the set of predictor variables does not include all those originally used. In this case there are no empty regions, but the sample is not a random sample from the overall population—it is distorted. We consider each situation in turn.

For classifiers based directly on the estimated posterior probabilities of class membership given measurement vector $\mathbf{x}$, it is obvious that those regions of the space which do not contain information about the $f(j|\mathbf{x})$ cannot contribute to an improved classification rule. That is, information about applicants falling in the reject region (their position in that region) is not helpful in developing a new rule. Of course, we can still extrapolate the old rule over the reject region to obtain estimated probabilities of class membership for applicants falling in this region. In fact, a popular practice in the credit industry is to use these predicted probabilities, weighted according to the distribution of applicants in the reject region, and combined with the known distributions and classes in the accept region, as the basis for an iterative technique to produce a new classification rule. This practice is misguided. It has no theoretical basis and there is no reason why it should lead to an improved classification rule.

Similarly, for classifiers which derive these posterior probabilities via Bayes rule from the class conditional distributions, we do not know their true class in the reject region and so cannot use points in these regions to estimate the $f(\mathbf{x}|j)$. Thus it seems that rejected applicants are also not helpful in constructing a new rule. Of course, just as for rules based directly on the $f(j|\mathbf{x})$, one could simply use the applicants in the accept regions to estimate the $f(\mathbf{x}|j)$, derive estimates of the $f(j|\mathbf{x})$ from them, and extrapolate over the reject region (or extrapolate the $f(\mathbf{x}|j)$ and then compute estimates of the $f(j|\mathbf{x})$). Whether this is feasible or not will depend on the estimation method used. It is unlikely to be helpful with nonparametric kernel or nearest neighbour methods, but would be helpful if global parametric forms were assumed for the distributions. Even then, one would need to be very wary of the sampling distortion caused by the censoring of the distributions at the decision surface. Although the original distributions may have been normal (improbable in this application, but just for the sake of argument), the distributions terminated by the hard decision surface boundary certainly would not have been. Such censoring would need to be allowed for in the estimation procedure; it would not be sufficient, for example, simply to use the straightforward sample means and covariance matrices.

It seems, then, that the information which is available about applicants in the rejection regions—that is, the information on their application forms (their position in the measurement space)—cannot be used with either type of classification rule to develop an improved classifier.

There is, however, one ray of hope. For classification rules of the second type—those based on indirect estimation via the $f(\mathbf{x}|j)$—*if global parametric forms are assumed for these class conditional distributions* then unclassified observations can help. In principle it is then possible to estimate the parameters of the component distributions from an entirely unclassified sample of applicants, using mixture decomposition methods. Such approaches typically require large samples, but in credit scoring applications that is usually no problem. More problematic is the need to assume that the classes (goods and bads in this application) have measurement vectors following a distribution from some family. The data in such applications is typically categorical and there is no obvious choice of family.

We now turn to the situation when the new predictor variables do not include all the information used in selecting the design sample. One might reasonably expect some design set points to occur in all regions of the new predictor space, though not proportionally to the original population distribution over the predictor space. This situation corresponds to the *nonignorably missing* case described by Little and Rubin (1987). In this case to obtain an effective classifier one will need to model the distortion induced by the original selection process. If information is available about the original selection mechanism (perhaps details are known about the original credit scoring instrument), then this can be used. If not, a formal statistical model of the selection process must be built. This latter situation is described by Copas and Li (1997). More general descriptions of reject inference are given in Hand and Henley (1993, 1994).

9.6 COMBINING CLASSIFIERS

As will have been made abundantly clear by the preceding chapters, there are many different methods for constructing classification rules. Chapter 11 compares these various methods, describing their relationships and pointing out the merits and demerits of different methods in different circumstances. The sorts of factors which influence performance are whether the data are continuous or categorical, the dimensionality, the complexity of the shape of the probability distributions and the decision surface, the aim of the classification, the speed with which a classification must be reached and the number of classes involved. In general, it is fair to say that no one method is universally the 'best'. This being the case, various researchers have considered combining classifiers, in the hope that a combination can be found which will yield final classifications superior to those of each constituent. Note that one way of looking at such multipart classifiers is to regard each constituent as a feature extractor. We return to this observation below.

Such combinations can be made in various ways, but a useful distinction is between those which (i) choose 'the best' of the constituent classifiers for each new object and (ii) those which combine the predictions of each of the constituents. We look at each of these types in turn.

9.6.1   Choosing 'the best' for x

Combination rules which base their classification on a choice from the set of constituent rules may have the advantage that they avoid computing classifications for all constituent rules. This could matter if speed is at a premium. The basic idea behind such rules is that the measurement space is partitioned into regions, each one associated with a constituent rule which best suits it. Such approaches are multilevel classifiers; the space is first partitioned into regions predicting which classifier is most accurate and then each region is partitioned into subregions predicting classes. Which classifier should be used for each region is, of course, determined from the training set. Given the multilevel nature, there are endless possibilities for combining classifiers, and there is enormous scope for consuming vast amounts of computer resources.

One obvious way to do it is implicitly by defining a local region around the point x to be classified and computing some performance criterion from the design set over this region for each classifier. Local regions can be defined as hyperspheres using some suitable metric and fixed radius, although this risks having few points fall within the hypersphere. This can be overcome by using a (kernel) function which decays smoothly with distance from x. Alternatively a nearest neighbour approach can be adopted, in which the radius of the region is defined as the distance to the kth nearest design set point. Another approach which has been used applies a 'gating' neural network to define the regions and to choose between the constituent classifiers.

9.6.2   Combining the predictions of constituent classifiers

The approach we have just outlined can be regarded as a way of weighting the constituent classifiers so that all but one of them has weight zero at x. This leads to the obvious generalisation in which all classifiers make some nonzero contribution to the decision. There are various ways in which this can be done. Statisticians tend to think in terms of weighted combinations and machine learning experts in terms of voting procedures (though these are generalisations, and exceptions can be readily found).

If the constituent classification rules have produced probabilities, then these can be averaged. For a weighted combination, weights need to be defined. Suppose that the error of classification rule j at x is $e_j(\mathbf{x}) = \hat{f}_j(1|\mathbf{x}) - f(1|\mathbf{x})$

(assuming two classes for simplicity). Then the expected squared error of a weighted sum with weights w_j is

$$E\left[\left\{\sum_j w_j e_j(\mathbf{x})\right\}^2\right] = \int \left\{\sum_j w_j e_j(\mathbf{x})\right\}^2 f(\mathbf{x})d\mathbf{x}$$

With a little standard algebra this can be rewritten as $w'\mathbf{C}w + (w'E(e))^2$, where $\mathbf{C}$ is the covariance matrix of the errors. If $E(e) = 0$ then (imposing the constraint $w'w = 1$) the set of weights which minimise this are the components of the eigenvector of $\mathbf{C}$ corresponding to the smallest eigenvalue. These ideas can be applied more generally, and do not require the e to be defined in terms of differences between probabilities and their estimates. They could, for example, be in terms of error rates.

More generally, one could again use values related to estimated accuracy of classification in the region of $\mathbf{x}$. Also, there has been substantial work by Bayesian statisticians on ways of combining the opinions of experts. If such opinions are expressed in probabilistic terms, these ideas translate directly into suggestions for combining classifier outputs.

Voting procedures can be on a majority basis or may require that agreement is unanimous, for example. The requirement of unanimous agreement leads to the reject option.

Combining classifiers into weighted combinations has an obviously Bayesian flavour, and one can approach the problem from that perspective. The obvious difficulty lies in choosing a suitable prior distribution over models. One approach is via *minimum description length* or *minimum message length*. This approach seems to have been followed in particular for tree classifiers. One of the problems (and not just for trees) is that the potential set of models is vast. On grounds of practicability this needs to be reduced (that is, some models need to be given a weight of zero in the sum). Various methods to do this using tree classifiers were described in Section 4.4, including path sets, options sets and fanned trees.

Simple combinations such as voting or weighted sums are straightforward ways of combining the predictions of different classifiers. However, recalling that the output of each constituent classifier can be regarded as a feature, an obvious alternative would be to use these outputs as inputs to a more elaborate classification rule. Although weighted sums yield a linear decision surface in the space spanned by the outputs of the constituent classifiers, more flexible surfaces can be used. This leads to the notion of multilayer classifiers, in which (layer 0) each constituent classifier produces a prediction; these predictions serve as inputs to (layer 1) a further set of classifiers (which could be of the same or different form as those in layer 0); the outputs of layer 1 serve as inputs to another layer, and so on. The parallel to multilayer feedforward neural networks is obvious.

To construct such a multilayer system one needs to induce a design set in the higher-level layers. One way to do this, illustrated for layer 1, is as follows.

Suppose there are N constituent classifiers. Take each of the n design set elements in turn. For each one, produce N predictions for its class membership, based on the other $(n - 1)$ points in the design set (as in the leaving-one-out error rate estimation method). These predictions serve to locate a point in the space spanned by the N predictions. The true class of this point is the true class of the corresponding design set element. These n points thus form the layer 1 design set.

9.6.3 Further reading on combining classifiers

Xu et al. (1992) illustrate various voting approaches to combining classifiers applied to handwriting recognition. Ho et al. (1994) use logistic regression to choose the weights with which to combine the classifiers. Mandler and Schurman (1988) and Xu et al. (1992) describe weighting methods based on the Dempster–Shafer theory of evidence. Huang and Suen (1995) describe a method for defining partitions, applied to recognising handwritten numerals.

Wolpert (1992) describes multilayer classifiers (he describes them as 'stacked') and Henery (1996) gives an illustration. Schaffer (1994) takes the idea further by letting the features obtained from the constituent classifiers be combined with the original variables to span a higher-dimensional space. Jacobs et al. (1991) describe the use of a neural network to partition the measurement space and hence to choose between lower-level neural networks. Work on combining neural networks using weighted sums of predictions is described in Perrone and Cooper (1993) and Perrone (1994).

Buntine (1992), Clark and Pregibon (1992), Quinlan (1992) and Oliver and Hand (1996) describe approaches to averaging over trees. Draper (1995) puts the model-averaging issue in a more general perspective than classification problems. A review of work on the combination of probabilistic opinions is given in Genest and McConway (1990) and a list of most of the key references in Bernardo and Smith (1994, pp.103–104). Minimum description length ideas are described in Rissanen (1987, 1989) and minimum message length ideas in Wallace and Freeman (1987).

9.7   CATEGORICAL DATA

Although a great deal of statistical theory is based on the assumption that data are real numbers, in fact all data are rational. Worse, all data are categorical, with a limited number of categories depending on the accuracy with which the values are measured and recorded. There are various implications of this. For example, no data ever follow a normal distribution and ties are inevitable for large enough data sets. One might also wonder about the impact on the accuracy of conclusions of the assumption that data can take any values within certain ranges when this is not the case. Formal approaches to this issue have been developed for some situations, such as assuming an underlying continuum which is partitioned by the measurement process, grouping observations together as in a histogram, and including the grouped data in the likelihood analysis (Heitjan, 1989; Heitjan and Rubin, 1991). However, it seems that, provided the grouping is fine, for most practical purposes the effect can be ignored.

In other situations, however, the effect must be taken into account. Sometimes the grouping is coarse, with only a few categories. Sometimes the groups have only an ordinal relationship, and it is not possible to give them relative magnitudes. Worse, sometimes only a nominal variable is involved, so the groups have no order.

An extreme case, the most common special case, arises when the variables are binary. One reason for its importance is that a common way of defining classes is in terms of complements: those objects possessing a property versus those not possessing it. Another is that categorical variables with more than two categories can be expressed as a set of binary variables, as explained below. Binary variables, scoring 0 and 1, can be viewed as a special case of both ordinal and nominal variables.

There are various approaches to analysing categorical data. One can, of course, ignore its categorical nature and treat it as if the observed values were obtained from a continuum. If the categories do have natural magnitudes associated with them (for example, if there is an underlying continuum which has been grouped to yield the values, such as age-groups), these magnitudes can be used directly in the analysis. On the other hand, if the scale is only ordinal, numerical values will have to be obtained before standard continuous data analyses can be undertaken. In medical and credit scoring applications a common assignment procedure is to use 'weights of evidence'. If there are no missing data, this is the logarithm of the number of class 1 objects in level i of the variable divided by the number of class 0 objects in level i of the variable.

Alternative weighting procedures can also be devised. At the extreme, one could regard finding suitable weights as part of the decision surface estimation procedure—an optimal scaling approach. In a sense, however, provided the classification rule methodology adopted is flexible enough, the precise weighting procedure used will not matter (given that it preserves ordinality, if necessary), so something as simple as ranking could be used. This is perhaps another reason why methods such as neural networks have attracted so much attention: they are flexible enough to cope with such data without requiring the user to think carefully about the variables.

Another alternative is to use indicator variables. The most straightforward approach, suitable with nominal variables, is simply to define a new binary variable for each level of the original nominal variable (except the last, which is determined by the others), coding these binary variables as 0 except when the corresponding level of the nominal variable is occupied, when they are coded as 1. Thus an object which takes the sth level of a nominal variable with r categories translates into $r - 1$ binary variables, all with the value 0 except the sth, which has value 0. (They are all 0 if the object takes the final value of the nominal variable.) Ordinal versions can also be constructed, in which variables take values 1 up to the sth level and 0 beyond that.

Classical multivariate statistical work is heavily based on multivariate normal distribution theory, so there is a natural tendency to view early statistical methods of supervised classification from the perspective of continuous data. It is only relatively recently that effective methods for categorical data have been developed. One reason for this is the lack of a categorical multivariate distribution playing a corresponding key analytic role to the multivariate normal distribution in the continuous case. The computer has also had a major impact here, since iterative estimation methods are often needed. Having said that, over the past two decades or so, statistical methods for categorical data have blossomed and some very powerful methods have been developed.

In contrast to the statistical community, researchers in computer science, artificial intelligence and pattern recognition have perhaps regarded categorical data as the norm. Various reasons have been suggested for this. One is the suggestion that statisticians tend to deal with natural universes, whereas computer scientists tend to deal with artificial universes. Another is the influence of database theory in computer science, in which variables are often fundamentally categorical. A third is the very fact that computer scientists are more familiar with the essentially digital nature of numerical encoding in modern computer technology, where everything is discrete. A fourth is the emphasis on perfect separability in the computing literature on supervised classification. With categorical data the space is partitioned into cells, so that separability follows more naturally than it does when continua are involved.

Once categorical data have been numerically encoded they can be analysed by any method. However, some methods might be expected to perform poorly in some situations. Take, for example, classical linear discriminant analysis applied with binary data. For simplicity, suppose that we have only two classes and that, initially, they are measured on a single binary variable. Let the probability that this variable takes the value 1 in class i be p_i. Then p_i is the mean for class i. Also, the variance of this variable in class i is $p_i(1 - p_i)$. This means that, provided $p_1 \neq p_2$ and $p_1 \neq 1 - p_2$, the classes must have different variances. Generalising this to more than two variables (all still binary), this means that, provided the mean vectors for the two classes are not equal (or can be made equal by interchanging 0/1 scores for some variables for one of the classes), the covariance matrices cannot be equal. Since classical linear discriminant analysis is based on an assumption that the classes have different mean vectors but the same covariance matrix, one might expect the method to perform poorly with multivariate binary data. Experiment suggests, however, that this is not always the case. (Though clearly this will be a complicated function of design sample size, accuracy of performance assessment, overfitting considerations, and so on.) In general, it seems that if the true decision surface is roughly linear then linear discriminant analysis performs reasonably well.

Numerical encoding, followed by regarding the data values as if they were scores on continuous numerical scales is one approach to developing classification rules for categorical data. But many other methods more naturally suited to categorical data have been developed. As with continuous data, some of these methods focus on the class conditional distributions, estimating each of the $f(\mathbf{x}|j)$ and combining them with the class priors to yield a classification, whereas others go directly for the $f(j|\mathbf{x})$ or even more directly for their differences or ratios. Where the categorical data approaches differ from the continuous data approaches is in the nature of the models constructed for these distributions.

An obvious approach is to construct the direct maximum likelihood estimators for the distributions, based simply on the number of observations falling within each cell of the cross-classification of the variables. This is only feasible when the *measurement complexity*, the number of cells in the cross-classification, is small. This will only be the case when there are few variables each with few categories. Without this being the case, an astronomic number of design set points may be required for reliable probability estimates. (Having said that, situations do occur when only a restricted set of the possible cells occurs or is likely to occur; then the straightforward maximum likelihood approach may be feasible. Again it is a case of matching the method to the problem.)

There are two basic approaches to overcoming this problem, as outlined in Chapter 1. One is to smooth the distribution and the other is to assume some kind of model form, which may be specified using only a relatively small number of parameters. A common smoothing approach uses the kernel method. Its use with categorical data is described towards the end of Section 5.2. Nearest neighbours methods—based on the numbers of points in each class which fall among the k nearest neighbour to the point to be classified—can also be extended to categorical data. A slight complication here is that, by virtue of the discrete nature of the measurement space, there may be several points at each possible distance from the point at which a classification is required (with d binary variables, the only distances which can occur are $0, 1, 2, \ldots, d$) so that the k nearest neighbours may not be uniquely defined.

Turning to modelling approaches, we find that many such approaches have been developed for multivariate binary data. Many of them expand distributions defined over the 2^d-point measurement space in terms of orthonormal basis functions, and then choose just low-order functions via some modelling process based on the design set. Some methods work with functions of the distributions, rather than the distributions directly. An overview of these sort of approaches is given in Chapter 5 of Hand (1981a).

Nowadays perhaps the most important modelling approach for categorical data is based on the log-linear model. This expands the logarithm of the expected values of the numbers in the cells in terms of a linear sum of terms which can be interpreted as main effects and interactions. Such models are described in Bishop *et al.* (1975) and in works on generalised linear models such as McCullagh and

Nelder (1989). An important subset of them is the class of graphical models (e.g. Whittaker, 1990).

Of course, multivariate data do not arise merely as all categorical or all continuous; both types of variables often occur together. A common approach then is to transform the variables so they are all of one type. One might, for example, numerically encode the categorical variables, as described above, or one might categorise the continuous variables. Either way, one can then use standard techniques.

An alternative is to develop methods which can work with both types of data simultaneously. One class of models developed specifically for this situation is the *location model* (Chang and Afifi, 1974; Krzanowski, 1975, 1983b, 1986). This assumes that the continuous variables have multivariate normal distributions with differing mean vectors within each of the cells defined by the cross-classification of the categorical variables, but assumes that the continuous variables have a common covariance matrix in the cells.

Recursive partitioning methods are well suited to handling mixed variable types. Such methods consist of nested partitions of the measurement space, with the subcells of a cell being defined in terms of a partition of a measured variable. With ordered variables, both continuous and categorical, the partitions of the variables are easily defined by a threshold. The methods handle both types of variable with equal facility. Unordered variables require an explicit grouping of the levels, but there is no difficulty in this.

CHAPTER 10

Some Illustrative Applications

10.1   INTRODUCTION

As has been emphasised throughout this book, different application domains lead to different problems, requiring different solutions. This is why it is not possible to identify the single 'best' classification method which can be used in all applications. Instead, it is necessary to be aware of the strengths and weaknesses of the different approaches, so that methods and applications can be appropriately matched and methods can be properly tuned. To give a flavour of the different issues presented by different application domains, this chapter briefly examines some widely differing classification problems.

10.2   AUTOMATED CHROMOSOME CLASSIFICATION

Chromosome analysis is used extensively in medical research and practice, especially in areas such as antenatal screening and tumour diagnosis. Carothers and Piper (1994) remark that chromosome abnormalities are present 'in perhaps 20% of all conceptions, in 50% of early spontaneous abortions, in 10% of mentally retarded individuals, and in many, perhaps most, cancers.'

Human cells have 46 chromosomes, grouped into 22 autosomal and two sex classes. Abnormalities can occur in either their number or their structure and it is these abnormalities that chromosome analysis seeks to detect. However, complications arise from (i) the fact that the preparation of cells for analysis often leads to chromosomes being hidden or lost so that it is insufficient to examine a single cell, and (ii) the fact that chromosomes are most easily analysed when the cells are in metaphase (dividing) and that such cells are relatively rare, so that many cells may need to be examined to find suitable ones.

As with all classification problems, a crucial aspect of chromosome analysis is deciding what to measure. Cells are prepared for analysis by a staining process which gives the chromosomes characteristic patterns of light and dark bands orthogonal to their major axis. Automated analysis methods are then based on features derived from this banding structure as well as general shape features. In particular, many features are based on width and density profiles along the length of the chromosome. Examples of these are the centromeric index (taken as the point of minimum width); the first few Fourier components of the density profile; other integral transformations aimed at producing particularly relevant features (e.g Granum *et al.* 1981); the height, width and position of components of a normal mixture distribution; for each peak of the density profile, the location, height and difference in height to the neighbouring valley; and Markov approaches summarising the probabilities of neighbouring bands occurring (e.g. Granum and Thomason, 1990).

These various possibilities have their merits and demerits. For example, Fourier decomposition produces overall summaries and so will miss key local features such as displaced bands; normal mixture components are all very well, but there is no notion of underlying populations from which the curves are drawn, so that normality is an arbitrary choice.

As with other classification applications involving longitudinal patterns, such as speech recognition and signature verification, there are problems arising from overall length distortions (due, for example, to different metaphases). Other distortions typically arise due to differential intensity of staining of the chromosomes. Some kind of normalisation is needed to compensate for these distortions. Although, here, we are looking at the feature extraction and classification problems separately, this is a rather unrealistic simplification of best practice. In fact, the classification technique adopted will depend on the features used. Moreover, in the present context, the extent and kind of the normalisation will depend on how far the representations of the chromosomes are from the representations used in constructing the classification rule; this cannot be determined without knowing the true classes. Too much 'normalisation' and the classification is to the wrong class. Worse, one could end up normalising an abnormal chromosome to the extent that it appeared to match a normal class. Iterative techniques have been explored to resolve this problem (e.g. Hildritch and Rutovitz, 1972)

The chromosome classification problem has a property common to several others we have considered in this book, namely that the classification of each chromosome is not independent of the others. This is because there are only 24 chromosome classes in a (healthy) cell, and exactly two chromosomes must be assigned to each class. For a particular chromosome, rather than choosing its class by maximising the posterior probability that it belongs to each class, we should maximise the overall likelihood of the distribution of chromosomes across the classes subject to the constraint of two in each of the 24 classes. (Other problems with this sort of constraint arise in credit scoring, tax investigations and medical screening.) In principle this can be done by examining all possible allocations satisfying the constraint, but in practice this is infeasible since there are too many possibilities. Various ways of searching through a subspace of the entire space have been explored, including: focusing an exhaustive search on those classes especially likely to be confused; transition models, which take an unconstrained allocation and then move individual chromosomes to new classes to satisfy the constraint; and recasting it as a linear programming problem.

A second constraint also applies to the chromosome classification problem, namely the fact that it is often reasonable to assume that all the cells in a sample have the same chromosome constitution. Whatever one's confidence about the classification for a single cell, one can have great confidence about the overall results from several cells. Indeed, one can even go further and use partial information from each of a number of cells, perhaps using only that information about which one is most confident.

A wide range of classification methods have been explored for automated chromosome classification, including classical methods based on assumptions of ellipsoidal distributions (Granum, 1982), recursive partitioning methods (Shepherd et al. 1987), and nearest neighbour methods (Lundsteen et al., 1981; Timmers, 1987). Finally, Carothers and Piper (1994) present an excellent overview of computer-aided human chromosome classification.

10.3   CREDIT SCORING

The term *credit* is used to describe the loan of an amount of money to a consumer by a financial institution for a fixed period of time. In such a transaction, the lender wants to be as confident as possible that the money will be repaid in due course (normally in parts, regularly at fixed times, and including interest). Also (fraud aside) borrowers would not want to borrow money if there were little chance of them being able to repay it. Thus there is a need to distinguish between good risk and bad risk applicants for credit, both from the lender's and the borrower's perspectives.

In reality there are not simply two well-defined classes of good and bad risks (see Section 9.4). A more accurate model might assign each applicant a probability of defaulting on repayment. We shall later mention a step in this direction, where an *intermediate* class is defined in addition to the good and bad classes. In any case, it is likely that the propensity to default will change over time for most people as external circumstances change. For example, losing one's job may make it very difficult to continue to meet mortgage repayments. There are also complex issues of what is meant by 'default'. Is it failure to meet a single one of the scheduled repayments on time? Is it failure to meet several successive repayments? Is it being behind schedule on several successive repayments? Should, in fact, default be measured on a multivariate continuum, in which one tries to capture more than one aspect? And so on. In any case, good and bad risk are arguably not of central interest to the lenders: what they really want to know is whether an applicant will be profitable for the company. Defaulters can be profitable if sufficient money has been repaid before the default occurs, and nondefaulters (good risks) who pay off their credit card bills each month are not profitable for the company. Profitability is a complex issue, and one which has not yet been adequately explored in the credit industry. Apart from risk probability, other factors which need to be taken into consideration in assessing profitability include the cost of collecting information, the cost of analysing it, expected returns on good and bad loans that have been accepted, the probability that someone offered a loan will decline the offer, and interest rates.

However, all these subtleties aside, the fact is that most work in the area has focused on identifying good and bad risk classes. Until a decade or two ago, classification into these two classes was made by human judgement on the basis of past experience. However, several factors have conspired to change this. One is the sheer scale of the current credit industry. In 1995 the amount owed to banks, credit card companies, building societies, retailers, mail-order companies and other lenders was of the order of £500 billion. Developing personal familiarity to permit reliable estimation of risk in all these loans is infeasible. Secondly, the tremendous advance in computer technology along with the development of the classification methods described in this book, made automated alternatives practicable. And thirdly, legislation aimed at preventing subjective prejudice from influencing decisions made the replacement of the human in the decision process much more attractive. This is apart from any increase in accuracy which may also result from automating the decisions (see, for example, the references cited in Rosenberg and Gleit, 1994).

In the industry, objective statistical methods of allocating individuals to risk classes are known as *credit scoring methods*. This term derives from summated rating scale methods, still very widespread in the industry. Each response on an application form is assigned a value and the sum of these values for an individual is that individual's overall score. This is compared with a threshold to produce a classification. Such a score is an *application score* since it measures the propensity of a new applicant to default. It can be contrasted with *behavioural scoring*, which is a score based on existing borrowers' repayment behaviour, and which can be used for such things as deciding what kind of action to take to pursue a mildly delinquent loan, deciding whether to offer a borrower a new loan, and so on.

The credit industry has also developed other terminology, rather different from standard statistical terminology (as seems to occur quite often in application domains). For example, what statisticians would call the variables, people in the credit industry call *characteristics*; and where statisticians would describe these as having different levels, in the credit industry they have different *attributes*. We shall stick to the statistical terminology here.

Rules for classifying applicants into good or bad classes will be developed from a design set, as usual. However, complications can arise in this application. Only rarely will the design set provide data on the entire measurement space. Typically, the information available for design will consist of measurement vectors for all applicants (which can be regarded as a sample from the population of future applicants) plus the true good/bad classes for only those applicants who were classified as good risks by some earlier classifier. Only these applicants will have been accepted for loans and therefore followed up to see if they are good or bad. This clearly presents great problems since information is missing at precisely those parts of the measurement space where it is needed (assuming the original classifier was not hopeless). This issue is discussed in more detail in Section 9.5.

Although we have stressed the objective of allocating applicants into two classes, good and bad, a common practice in the industry is to define a third, intermediate, class, but to use only the good and bad design samples to develop a new scorecard. It is difficult to justify this on statistical grounds: a more accurate classification rule would result if the intermediates were included in some way since it is precisely in the 'intermediate' zone where most misclassifications will occur. Hand and Henley (1996a) discuss this issue further. Classifying applicants as intermediate leads to an obvious application for the notion of the reject option, in which a decision is deferred about applicants for whom a confident classification cannot be made (perhaps while more data is collected). Indeed in some credit applications precisely this practice is followed. For example, in mail order a short application form printed in a magazine is sometimes followed by a more detailed one for people whose score on the short form is in a middle zone.

Data sets in credit applications are typically large; there may be tens of thousands or even millions of cases. This means that overfitting is often not a problem, and that sophisticated methods of error rate estimation, such as leaving-one-out and bootstrap methods, may not need to be employed. On the other hand, since the information in such data is commercially sensitive, access may be restricted. Missing data is often a problem. Most of the variables are categorical, and if not, the practice is usually to categorise them prior to analysis. However, having done that, it is most common to numerically code the categories and use these numbers in subsequent analysis. Table 10.1 gives some examples of variables.

Numerical codes could be avoided by using dummy variables or could be found by using optimal scaling methods, but a widespread practice in the industry seems to be to use weights of evidence: the jth attribute of the ith

Table 10.1   Characteristics typical of certain credit scoring domains.

Time at present address (years)	0–1, 1–2, 3–4, 5+
Home status	Owner, tenant, other
Postcode	Band A, B, C, D, E
Telephone	Yes, no
Applicant's annual income (£)	£0–10 k, $11 - 20k$, $21k+$
Credit card	Yes, no
Type of bank account	Cheque and/or savings, none
Age (years)	18–25, 26–40, 41–55, 56+

characteristic is scored as $w_{ij} = \ln(p_{ij}/q_{ij})$, where p_{ij} is the number of goods in attribute j of characteristic i divided by the total number of goods (who respond to characteristic i) and q_{ij} is the number of bads in attribute j of characteristic i divided by the total number of bads (who respond to characteristic i).

Credit data sets often have many variables—over a hundred responses for each applicant is not unusual—but this will obviously depend on the area (less information will be sought for someone wanting to buy a £30 toaster on hire purchase than will be sought for someone wanting to take out a half million pound mortgage). Given the typical sizes of the data sets it will be advantageous to use as many variables in the classification rule as possible—overfitting is not likely to occur. On the other hand, too many questions and too searching an application procedure may deter potential borrowers.

A further property of the data which can cause difficulties with some types of classification procedure is that the variables are often structured. That is, some questions may only be asked contingent on certain answers to earlier questions. This means that, for example, classical statistical methods such as linear discriminant analysis cannot be applied without modification.

Population drift—the population changing over time—is also a potential difficulty here, where the behaviour of borrowers is influenced by short-term pressures (such as Budget announcements by the Chancellor of the Exchequer). This means that speed of developing *and implementing* classification rules is important, and they will probably need to be changed quite frequently (depending on the precise application area). It also means that there are advantages to methods which can be dynamically updated, without requiring recomputation from scratch. For example, the nearest neighbour method can be arranged so that the oldest design set elements are dropped and new (classified) applicants added as the system is used.

Depending on the classification method adopted, population drift will not normally be a problem if the conditional probabilities of belonging to each class given the measurement vectors remain unaltered, even though the overall distributions evolve. This means that, if essentially all of the possible predictive power in distinguishing between goods and bads is contained in the variables used, then population changes will not affect performance. On the other hand, if powerful predictive variables are missing from the model,the conditional probabilities are likely to change as the overall population changes.

A common criterion of performance in the credit industry is the proportion of those applicants granted a loan who then default. The underlying motivation is illustrated by a bank which might have £X to lend—equivalent to N loans—and obviously wants to minimise the number of defaulters. The good risks who have been rejected, although important from the perspective of the bank's image, are less important for short-term profits. This means that error rate, combining the two types of misclassification, is of limited importance.

We have already noted that in the credit industry the most common classification technique is a comparison of a summated rating scale with a threshold.

Since, however, small improvements in classification accuracy can translate into large improvements in financial terms, there has been great interest in alternative classification methods. In particular, whenever apparently new methods arouse interest (such as recursive partitioning methods a few years ago and neural networks more recently), credit lenders are quick to explore their possibilities. Extent of usage in the industry may therefore be adopted as a measure of effectiveness. It is thus a salutary lesson to note that, despite all the theorising and the hype over other methods, summated rating scales are still the most widely used.

Simple classification accuracy (even if measured by proportion of bad risks amongst the accepts, rather than error rate) is but one aspect of performance. Others, also important in this context, are the intuitive sense underlying the scores (for example, if the levels of a variable are ordered, it is attractive that the numbers assigned to those levels should have the same order) and the ease with which an explanation can be given for a decision (in some countries there is a legal obligation to tell a rejected applicant why they were rejected, if asked).

The numbers to be assigned to each level of each variable are most commonly found using regression or discriminant analysis. Both have the merit of being relatively conceptually straightforward and widely available in statistical software packages. The coefficients from such analyses are combined with the weights of evidence and rounded to whole integers (after rescaling, if necessary) to give the values to be added to yield the overall score.

The first published account of discriminant analysis in the credit scoring context seems to be that of Durand (1941). Of course the categorical nature of the data, and the lack of any reason why the covariance matrices of the good and bad classes might be supposed to be equal, might lead one to be wary of using discriminant analysis. However, Reichert et al. (1983), based on empirical study of credit scoring problems, concluded that 'the fact that a significant portion of credit information is not normally distributed may not be a critical limitation.'

On theoretical grounds, one might suppose that logistic regression was a more appropriate tool than linear regression in this context, but practical experience (e.g. Henley, 1995) suggests that there is little to choose between them. This might be because most of the applicants fall in regions of the measurement space where the probability of being good lies between 0.2 and 0.8. In such regions the logistic surface is approximately linear.

Classification trees are used in this application, but often in a concealed way, rather than to produce the overall classification. In particular, they occur as what are sometimes called *derogatory trees*: small classification trees used as additional variables in a linear classification rule. That is, they are used to encode interactions.

Nonparametric methods, especially nearest neighbour methods, have been applied in several published studies. For example, Henley and Hand (1996) describe a detailed investigation of such methods applied to data from a large mail-order company, with particular emphasis on the choice of metric used. The attractive potential of such methods for dynamic updating has already been mentioned.

Neural networks have been explored, though actual everyday implementations seem to be limited. Rosenberg and Gleit (1994) describe several applications to corporate credit decisions and fraud detection and R. H.Davis et al. (1992) compare such methods with other classifiers for credit scoring.

Finally, remember that credit scoring is a commercially sensitive application. Banks are not in the business of publishing scientific papers, least of all papers which would reveal the basis of any commercial edge they may have gained. So one can only speculate on the results of work carried out behind the scenes. Having said that, the industry is a volatile one (at present, at least) with fairly rapid staff movements, so that any significant technical advance would rapidly percolate through the industry and would become generally known quickly.

10.4   SPEECH RECOGNITION

Speech recognition is one of the oldest application areas for the ideas of statistical pattern recognition and machine learning. Work dates from at least the 1950s (for example, K. H.Davis et al., 1952; Olson and Belar, 1956; Forgie and Forgie, 1959). It is also one of the areas which has attracted the most funding. The problem has proved a difficult one, but a huge amount of research has led to viable systems now being available and in everyday use. Indeed such sytems for isolated word recognition using a limited vocabulary (such as 'yes' and 'no' and the digits) to lead a user down a classification tree to a terminal action node have been available since the 1970s. Systems now exist which have effectively zero error rates for the 10 isolated digits with arbitrary speakers and less than 5% error rates on vocabularies of over 1000 words and a handful of specified speakers.

A speech signal can be represented in either the time or the frequency domain. The time representation shows just a single waveform evolving over time. Different parts of this signal will show silence, voiced sounds (when the vocal cords are vibrating, as in producing vowel sounds) and unvoiced sounds (when the vocal cords are not vibrating, as in sibilants). The aim is to find where to split the signal and to match the component parts up to the spoken sounds. The frequency representation takes short time slots (for example, 15 ms slots, shifted by 1 ms) and computes a spectrogram for each slot. These can then be displayed as a two-dimensional diagram, with time on the horizontal axis and frequency on the vertical axis.

A *phoneme* is a component sound of speech: the atoms from which the molecules (words) of speech are made. The number of different phonemes is rather a matter of convention, since the distinctions (between both the sounds and the waveforms) are not always clear-cut. A common decomposition is into around 50 phonemes, arranged in various classes, such as front vowels, mid vowels, back vowels, diphthongs, voiced and unvoiced stops, voiced and unvoiced fricatives, and so on. Vowels play an important role in speech recognition, but early work very rapidly led to the realisation that the same vowel sound produced by different speakers varied substantially, often with overlap of the different vowels. To cater for this, standardisation of some kind is required.

From the above description, an obvious approach to machine speech recognition is to decode the speech signal in a sequential manner as it arrives, identifying each phoneme signal as it presents. This requires, firstly, that the signal be segmented into intervals such that each interval corresponds to one phoneme, and secondly, that the words which the sequences of phonemes are making be recognised. At each time point phonemes are ranked according to their probability of occurring, based on a match between the signal and the typical characteristics of each phoneme. Words are then constructed by finding the sequence of phonemes which is most probable, subject to the constraint that it forms a word from the vocabulary being used. This constraint is similar to that arising in the contexts of automated chromosome classification, health screening, credit scoring and tax investigation (see Sections 8.2, 10.2 and 10.3). Although this sounds straightforward in principle, in practice it may not be easy to decide how the input signal should be partitioned into phonemes.

Whatever approach is adopted, the first step is to summarise the incoming signal in terms of descriptive parameters, or features, which can serve as input to the classification tool. Many feature sets have been proposed, involving highly complicated preprocessing stages, the details of which we cannot go into here. For example, in *filter bank analysis* the signal passes through a set of d bandpass filters covering the frequency range of the signal (each filter measures the energy of signal in its band so that a feature vector of length d emerges). In *linear predictive coding* the features are the set of weights in a model which predicts the current speech sample as a linear combination of previous speech samples. And in *cepstral analysis* the features are the coefficients of the Fourier transformation of the log magnitude spectrum.

Many classification methods have been used to assign the derived feature vectors to a class. They include tree-based methods, methods based on a flexible parametrised decision surface such as neural networks, and nonparametric methods such as nearest neighbour methods. However, there are complications. In particular, different presentations of the same word will differ in length and, worse, in the relative lengths of the constituent parts. Some way of aligning the presentations is needed. This problem is not unique to speech recognition; it also occurs in signature verification, for example. Typically, nonlinear transformations will be needed. These transformations, called *warps*, can be general, subject to certain constraints such as fixed end points, monotonicity, etc. Finding the optimal transformation—that yielding the best match between an incoming signal and a design set feature vector—is clearly a difficult task, and one that is well suited to dynamic programming techniques.

A popular classification technique in speech recognition is *vector quantisation* (Gersho and Gray, 1992), which takes the design set feature vectors for a particular phoneme and summarises them by relatively few prototypes, typically the centroids of some cluster analysis or mixture decomposition procedure. New feature vectors are classified on the basis of a simple nearest match to the prototypes for each phoneme. More sophisticated versions of this idea model a sequence of input vectors simultaneously, so that overlap between sounds can be coped with. Sometimes vector quantisation is used as a reject option preprocessor stage to eliminate unlikely candidates, and a more sophisticated classifier applied to the possibles.

The principle underlying vector quantisation is also sometimes applied at a higher level: sounds are grouped into subsets on the basis of their similarity. A new signal is first classified into a subset before closer examination matches it to a particular sound within that subset. This can lead to substantial saving in computation, increasing the speed of classification. It is essentially a tree-based classifier.

Much thought has been given to the question of what metric should be used to define similarity between two feature vectors in speech recognition. Mathematically tractable measures (such as straightforward Euclidean distance) may not have much relationship to the subjective similarity of two speech sounds. Research has shown what sort of differences between spectra correspond to large perceived differences between sounds and what sort of differences between spectra correspond to small perceived differences between sounds. For example, a large perceived difference is produced by a difference in the formant locations (a formant is one of the frequencies in speech carrying the most energy). And a small perceived difference results from a comparison between a spectrum and a notch filtered version of it, in which the signal is greatly attenuated in a narrow frequency band. For feature vectors based on filter bank analysis, distances based on (possibly covariance weighted) L_p-norms are often used, and for feature vectors based on linear predictive coding and cepstral analysis, likelihood-based measures and cepstral distance measures are often used. More sophisticated methods are based on studies exploring the nonlinear relationship between the subjective perception of the frequency content of sounds and their physical content. Such methods transform the frequency scale so that a physical spectrum can be converted into the subjective scale and apply a suitable distance measure to the result. Still more sophisticated methods recognise that speech is not a static process, implicit in the above, but is dynamic; they include dynamic features (such as a time differential log spectrum) in the recognition process. This last point is obviously related to the need for nonlinear transformation of the signal over time, mentioned above. Other methods take advantage of the fact that differences between phonetically similar sounds may occur only in a small part of the signal, and they weight this part accordingly in the comparisons.

Section 10.3 referred to the problem of population drift, in which the population of objects to be classified evolves over time. The same sort of phenomenon can arise in speech recognition, when new speakers are presented. Various approaches to this problem have been explored, such as analysing a standard input from the new speaker and using this to modify the stored prototypes, or adjusting the prototypes over time (as pointed out in Section 10.3, this strategy has also been suggested in the context of credit scoring). In general, systems often degrade dramatically in the presence of distortion or background noise. Such problems can include speech artifacts, such as heavy breathing or lip pops, and distortion arising from stress or other psychological conditions. The best solution is to train the system in the presence of characteristic noise or distortion, if this is possible. In some circumstances it is possible to use another input channel, which just contains the noise, and to subtract this from the combined speech and noise channel.

So far, the methods we have described have made no attempt to make a formal model of the processes underlying the signal. In fact, most applications of supervised classification seem to be like this. Just occasionally, as in the use of random effects models for analysing spectrographic data, explicit model structures are postulated and used. In speech recognition, one formal model structure which is widely used (and which is in fact related to random effects models) is the hidden Markov model (Ferguson, 1980; Rabiner, 1989). The essence of such models, making them appropriate for speech recognition, is the time-ordered nature of the signal. They are based on the notions underlying syntactic pattern recognition (e.g. Gonzalez and Thomason, 1978) and assume a number n of *states*, which cannot be directly observed. As time progresses, transitions between states occur, with unknown probabilities. Associated with each state there is a number of possible observations. Each of these possible observations will be observed with unknown probabilities. Finally, a distribution is assumed for the probability of initially being in each state.

Each spoken instance of a word is split into time intervals and the signal within each of these intervals is coded as that member of a finite set of spectral vectors to which it is most similar. Using the data from multiple instances, a hidden Markov model is constructed for each word. A new word is classified by running its partitioned signal through the models for each word and seeing which model has the highest probability of having generated the signal in question.

Isolated word recognition is all very well, but it is a very different problem from recognising continuous speech, the difference arising from the fact that the isolated words have well-defined start and end points. On the other hand, continuous speech signals contain linguistic knowledge not available in small-vocabulary, isolated-word problems. For example, syntactic knowledge leads to estimates for the probabilities that specific words will follow other words, so influencing the low-level classification. A final high-level semantic analysis can lead to rejection of meaningless sentences.

A characteristic of supervised classification applied in speech recognition is that there is considerable *structure* to the problem, and this structure can be utilised. We have seen this, for example, in hidden Markov models, modelling the temporal structure of the incoming signal, and in continuous speech recognition, where higher-level aspects of the relationships between speech components (such as grammatical components) are used to help determine the probability of different word strings. Constraints that lead to better models, and therefore more accurate probability estimates, will generally give a better performance in classification problems.

We noted at the start of this section that a vast amount of research and development has been carried out in the area of speech recognition. Two first-class reviews of the area are Rabiner and Juang (1993), an integrated and accessible textbook covering all the main issues in depth, and Waibel and Lee (1990), a large collection of the key papers in the area.

10.5 CHARACTER RECOGNITION

Speech recognition is one area where supervised classification techniques promise to have a major effect on everyday life, but character recognition has already made an impact; commercial optical character recognition systems have been readily available for some decades. In contrast to most of the structureless and context-free situations which characterise many pattern recognition problems, character recognition is characterised by the structure of the shapes being recognised and the relationships of their component parts.

The simplest method which has been explored is template matching, for example, the *peephole* method. Suppose that a particular character is consistently written in such a way that each pixel in the field is black or white—black where the strokes of the character occur and white elsewhere. Then a subset of pixels can be identified such that the black/white vector of these pixels is different for each character. (In the worst case the entire set of pixels covering the field could be used, but for reasons spelt out elsewhere in this book it is better to use far fewer.) The particular chosen set of pixels represent the 'peepholes'. The problem then reduces (in this ideal situation involving no noise) to one of simple pattern matching of the obtained peephole pixel vector with a dictionary of standard vectors. More generally, to allow for noise, a distance measure will be used to identify that standard vector most similar to the new one.

One of the problems with this sort of two-dimensional template matching is registration. A character not centrally located or perhaps incorrectly oriented could be misclassified. Either some kind of normalisation is called for, or methods which are suitably invariant are needed. Some approaches to normalisation are based on lower-order moments, perhaps normalising the position, size and slant of the characters. Sometimes higher-order moments are then used as features for a classification rule. In a sense, this represents a transformation of the raw character field into a truncated series expansion. Another obvious truncated expansion approach is via a Fourier series, which has been used in various ways; for example, using a circular harmonic expansion of the two-dimensional field or via Fourier summaries of a trace round the outer boundary of the character.

Peephole matching is simple, but it is not well suited to the problem of recognising handwritten characters because their variation is too great. What is invariant about handwritten characters, and what enables us to recognise them, is their structure: they have certain parts and these parts are related in certain ways. More sophisticated methods are based on deeper identification of these structures. For example, one class of approaches replaces the set of observed pixels by a set of narrow scans across the character, recording the number of times each scan crosses a black line. One might use several parallel scans made in each of several directions (e.g. horizontally, vertically and two diagonals). Alternatively, several key points in the field are sometimes identified and the scans are then radiated outward from these points.

Deeper still, aspects of the individual strokes of the characters are sometimes used as features, e.g. quantising the direction, length and position of each stroke; or the relative positions of endpoints, crossing points and branching points of a character, perhaps after a *thinning* or *skeletonisation* process in which the drawn or printed lines of the character are reduced to widthless idealised lines. Sometimes the stroke segments and relationships are represented as a mathematical graph structure, matched to elements of the design set using graph-matching algorithms. Some of these techniques have been explored in scene analysis. Taking this structural approach even further, syntactic pattern recognition methods scan the field in a particular order and label the features found in each subregion examined. The resulting string of labels then provides input to a parser which classifies it into an appropriate class. Methods of syntactic pattern recognition are discussed in Gonzalez and Thomason (1978) and issues arising from character recognition in Ali and Pavlidis (1977).

The complexity and the amount of computation which goes on behind the scenes in a practical system is nicely illustrated by the following description of the Hitachi H 8959 system, produced in the early 1970s (Mori *et al.*, 1992, p.1049):

'The machine at first employed a regular thinning algorithm. The thinned line was represented by chain encoding using 3×3 masks. Singular points, such as end points and branching points, were also identified by counting matched mask(s). If the number of matched masks was one, two, three, or four, then the end point, normal line point, branching point (T type), and intersection point (X type) were identified respectively. Thus the thinned character line is decomposed into branching line segments whose terminal points are singular except for a simple loop in which the starting point was used. The coordinates of those singular points of the branching line segments were registered. Then each branch segment was further divided into four kinds of monotone line segments, called four-quadrant coding: upward right and vertically up; upward left and horizontally left; down right and vertically down; and down left and horizontally right.'

If structure and context play a key role at the level of recognising individual characters, they can also play a role at a higher level. In particular, characters often occur in words or other structures (such as postcodes) with restricted syntactic forms, and this higher-level information can be used to constrain the matching process (cf. speech recognition) (Nagy, 1992). Taken to its extreme, this involves a hierachy of systems, each level informing those above and below it, with characters at the bottom level and a deep knowledge representation of the document at the top.

In this context, a word or two about recognising cursive script is also appropriate. The problems here differ from those of printed character recognition because the characters within a word are (typically) connected together. The distinction is the same as that between recognising isolated words and processing continuous speech. Two approaches have been used: one can either attempt to recognise each word as a whole, not bothering to identify the individual letters, or one can seek to segment the word into the letters. In signature verification, a dynamic approach is often adopted in which the time and pressure of each stroke are recorded and used as measurement space characteristics. This makes it much harder to forge a signature. These online approaches also have advantages for languages with a large alphabet of symbols, such as Japanese and Chinese.

Recent work explores combinations of classification rules, as outlined in Section 9.6. For example, in handwritten digit recognition Cohen *et al.* (1990) use (a) polynomial discriminants, (b) structural analysis of piecewise linear approximations of the contours of the digits, (c) structural analysis based on position and size of horizontal and vertical strokes and (d) a contour analysis method based on curvature. These four methods are combined using a decision tree. Suen *et al.* (1992) use a majority vote to combine the decisions of four classifiers, also in recognising handwritten digits.

Finally it is important to note that a major force in the practical commercialisation of optical character recognition systems has been technological progress in the scanning systems, particularly the introduction of the laser scanner in the early 1970s.

Mori *et al.* (1992) present a fascinating historical review of how the field has developed. They make the point that research into optical character recognition has been especially active in Japan and many important papers have not been translated into English. The special issue of *Proceedings of the IEEE* on optical character recognition (July 1992) provides an up-to-date outline of the state of the art.

CHAPTER 11

Links and Comparisons between Methods

11.1 INTRODUCTION

We began this book by asking, Which is the best classifier? And we suggested that the answer to this depended on two things: (i) the characteristics of the problem and (ii) what was meant by *best*.

Part II of the book was essentially structured in terms of the different methods. This meant that the discussion of the properties of the methods arose in the context of those methods. In fact, of course, only researchers concerned with methodological studies of classifiers will be concerned with this perspective. Researchers who want to apply the methods will want to approach things from the opposite direction; they will want a discussion of properties of classification rules in the context of the properties of their problems. This chapter attempts to present such a perspective, identifying important features of problems and providing a comparative assessment of classification methods in the context of those problems.

As far as *best* is concerned, this is often interpreted superficially as *classification accuracy*. But classification accuracy is just one aspect of performance. Others which may be just as important, perhaps even more so in certain contexts, include such things as:

- The facility with which an explanation for a classification can be given.
- The speed of classification, the shortest maximum classification time, the shortest average classification time, and so on.
- The cost of classifications—this may depend on what variables and how many variables need to be measured to reach a classification, how much computation is needed, etc.
- The amount of prior knowledge about the problem required to construct an effective rule.
- How readily the method can cope with missing values.
- How much user tuning is required. Are all parameters estimated automatically from the design data? Is a separate tuning set required?

- Any restrictions on the type of data which the method handles.
- Whether one is seeking to build a classifier which can be used in practice or merely to demonstrate the value of a method. This is related to whether one wishes to explore performance conditionally on a given design set (the problem typically encountered in practical applications), or whether one is trying to establish general performance levels (so one wants unconditional conclusions, which apply across a range of design sets).
- How easily the variables can be measured. Many classification problems are motivated by a desire to replace a costly/slow measurement process by a cheap/quick one. Loss of classification accuracy is often an accepted cost (though, naturally, one wants to minimise the loss by using the best method available). Essentially we are discussing a compromise between classification accuracy and other aspects of classification rule performance.
- Whether there are important constraints on the problem which the classification method must satisfy. In Section 10.2 we pointed out that the class sizes are fixed in chromosome classification, and this also applies in screening problems such as certain types of tax evasion and credit scoring tasks (though only probabilistically in these cases). Constraints also apply in continuous speech recognition, where the probabilities of occurrence of individual words are influenced by context in the form of higher-order syntactic and semantic structure of the language, and in spatial problems, such as oil exploration, where the class of a rock specimen from a drill core may be expected to be related to the classes of its neighbours.

There has been very little discussion of these more general aspects of performance and classification rule construction in the literature. Brodley and Smyth (1996) and Hand (1996) are exceptions.

Even if one has determined that classification accuracy is the important factor, careful definition is needed. Classification accuracy may mean error rate, expected overall cost, sensitivity, specificity, bad rate among accepts, Gini coefficient, precision, and so on—the various measures discussed in Part III. As we have seen, good performance with one measure need not necessarily mean good performance with another.

A general point is worth making here about classification accuracy. If classification rules have been developed in depth (by experts on the data, carefully applying existing methods on appropriately selected and transformed variables) for a particular problem domain, it is unlikely that a new type of rule will lead to substantial improvement in classification performance (though small improvements may be achievable). If large improvements *are* possible, they are more likely to come from a reformulation of the problem (such as by defining new class structures) or from the identification and inclusion of new variables. In any case, if the error rate (to use one illustrative measure of classification performance) for standard rules is already small (5% say), vast improvement is impossible, by definition. Of course, whether an additional 1% is worth achieving will depend on the problem. Hand and Henley (1996a) argue that small improvements obtained under 'laboratory conditions' may not transfer to real application environments.

In this concluding chapter we make a comparative critical assessment of the different methods described in Part II, attempting to show their strengths and weaknesses.

11.2 PRIOR KNOWLEDGE OF THE PROBLEM

One key feature which distinguishes between problems is the amount of prior knowledge the researcher has about the problem. Sometimes classification problems arise as one of a stream of similar problems, and sometimes they are effectively completely novel. An example of the former may occur in medical diagnosis, where perhaps data of the same sort, on similar diagnostic issues has been analysed many times before. Examples of novel problems are less easy to describe, for obvious reasons. Sometimes the data have been subjected to extensive analysis, in which distributional forms have been estimated and evaluated (outliers detected, multimodality recognised, etc.), before the classification rule is constructed.

The importance of prior knowledge arises because different classification methods are more or less flexible in the decision surfaces they fit. For example, classification trees, nearest neighbour methods and neural networks assume little prior knowledge of the shape of the decision surface and can fit complex surfaces. In contrast, the more 'traditional' statistical methods, such as logistic regression or methods based on covariance matrices of the predictor variables, fit more restricted forms. This does not mean they are necessarily less effective—indeed they can be more effective because any demerits due to extra bias may be compensated by the merit of less variance, as described in Chapter 1. What it does mean is that one needs to understand the problem and the data sufficiently well to know if and when transformations and combinations of the raw variables need to be used alongside them.

Essentially, what this is saying is that, to get the best from the less flexible methods, one needs to know that a decision surface of the kind they fit is appropriate for the raw variables or one needs to undertake a manual feature extraction process. This is in contrast to the more flexible methods such as neural networks, which integrate feature extraction with classification rule construction. They do the feature extraction for the users and require no expertise from them. (They therefore lead to complex rules, defying simple explanation; see the next section.)

The history of statistical pattern recognition is quite interesting in this context. In the 1960s and 1970s, much emphasis was put on feature extraction: how to find suitable combinations and transformations of the raw variables so as to enable an effective classifier to be produced. With the recent torrent of work on neural networks this emphasis has died down. In a sense, the hidden layers of neural networks provide automatic feature extraction stages so the system automatically decides how to combine the raw variables. Of course, one would expect the larger search spaces which this produces (since now it is necessary to search the space of possible combinations/transformations of variables as well as the space of decision surfaces) to lead to slower and more complicated algorithms, and this is exactly what one gets.

One of the reasons that neural networks have so caught the public imagination (regardless of their actual performance) is because they require a comparatively small understanding of the problem. An amateur can, in principle, obtain results as good as those of an expert. The machine does the work.

Of course, it is worth adding the cautionary note that nothing can beat sound theoretical knowledge of the data. If one knows, a priori, that the probability of belonging to class 1 is likely to be related to the square of variable x, then the square of x should be included as a predictor. Relying on the system to work out that x^2 should be included is unlikely to lead to as good a rule; after all, the system has to separate this truth from the superfluous random variation arising in the particular design set you have given it. As Minsky and Papert (1969, p. 16) put it, 'Significant learning at a significant rate presupposes some significant prior structure.'

11.3   POSTERIOR KNOWLEDGE OF THE PROBLEM

Although the emphasis in this book has been on classification performance, however that is measured, another possible objective is *understanding* of the data. This might be 'data analytic' understanding, meaning an appreciation of which variables turn out to be important in producing accurate classifications, which regions of the measurement space lead to clear-cut classifications, etc. (For example, Taylor and Silverman (1993) say, 'Our emphasis in using classification trees is on interpretation and exploration of a set of data.') Or it might be understanding for explanation; often it is necessary that a client or potential user has an appreciation of the methods and rationale behind a classification rule. In some contexts, such as credit scoring, this is a legal requirement; see Hand and Henley (1996a). In others, such as medicine, the system provides a 'second opinion', so it is important to understand the basis on which it has reached its conclusions. (In other applications, such as speech recognition, it is not important at all.) Logistic regression, classical linear discriminant analysis, belief networks and tree-based methods are effective in permitting simple explanations. Neural networks, nearest neighbour methods and kernel methods are less so. Methods based on complex mathematical or theoretical models may or may not be effective in this regard; they may, for example, not permit ready decomposition into explanation of which variables matter (not surprisingly, since typical problems are intrinsically multivariate, nevertheless this is sometimes required).

As far as tree classifiers go, one difference in emphasis between the statistical and machine learning communities is worth mentioning. While the statistical community will stop with the tree structure, the machine learning community will often use this structure as a basis from which to formulate the antecedent–consequent rules of an expert system. In principle, the conditions which have to be satisfied within a node in order to determine which offspring path is followed represent a rule's conditions (antecedents) and the outcome path represents its consequent. Because of this, the machine learning community places greater emphasis on the interpretability of each separate node.

Chiogna (1994) compared classification trees and belief networks for diagnosing congenital heart disease in newborn babies. The conclusion, perhaps not surprisingly, was that the trees yielded superior prediction but that the belief networks gave better understanding of the problem:

'Although it is expected that belief networks provide considerably more insight into the true structure of the problem, the results are penalised in terms of accuracy by the complexity of the relationships that the networks need to take into account, relationships which are not easily estimable from relatively small and complex data sets. On the other side, considering a belief network merely as a classificatory tool appears to be a restrictive use of a technique, whose role extends far beyond the solution of a classification task. Nonetheless, the evidence seems to suggest that in real and complex situations with limited data available, data-derived class-probability trees outperform data-derived belief networks in supporting a diagnostic task.'

Given the comments elsewhere in this book about the performance likely to be achieved by careful application of different classification rules, we expect these conclusions to apply to other types of classification methodology in comparison with techniques aimed at discovering the 'true' underlying structures relating the variables. Our overall conclusion is that one must decide whether prediction or understanding is the real objective.

As a prescription of a classification process to be followed by a human, classification trees have the merit of simplicity and ease of understanding and application. They require no calculations (at least, this is true of the simpler forms). They have found a wide variety of applications, for example, in fault diagnosis of machinery (such as computers and photocopiers), in medical diagnosis by paramedics, in more structured approaches to medical diagnosis and in botanical identification.

11.4 THE DATA

Some methods are more suited to different data types than others. The classical multivariate statistical methods perhaps sit more easily with continuous predictors than with categorical predictors. Of course, numerical assignments can be made to the categories of categorical predictors (by optimal scaling or by a simpler method, such as merely using weights of evidence) but this seems a rather ad hoc approach. An alternative is to use indicator variables, but this can lead to a huge number of raw variables. On the other hand, as we have noted elsewhere, even classical linear discriminant analysis often does surprisingly well with categorical data. Historically, the machine learning community has placed greater emphasis on categorical data than has the statistical community.

Tree-based methods can handle both continuous and categorical (and mixed) data with equal facility, and this is one of their attractions for real problems. The other 'nonparametric' methods, namely kernel and nearest neighbour methods, require suitable metrics to be defined. Tree classifiers are attractive because the basic forms of tree classifier are invariant to monotonic transformations of the variables and they are robust to outliers.

Missing data are a potential problem in any multivariate situation and supervised classification is no exception. Data may be missing from the design set or from the measurement vector of a new object to be classified, and these two situations require different solutions. Data may even be 'missing' deliberately. One example is personnel classification, where some questions may be contingent on particular answers to others (this situation arose in the osteoporosis data mentioned in Chapter 8). Another example arises when variables are measured sequentially and class membership probabilities are computed at each step; then classification is made when some confidence threshold is exceeded. An extension of the reject option, this idea means that the number of variables to be measured on a new case cannot be determined a priori. Of course, the standard form of the reject option may also be viewed as a case of missing data.

There are various ways to handle missing data in the design set. At the simplest level, it can be ignored—one merely analyses those cases which have complete measurement vectors. But it hardly needs pointing out that this risks bias. Questions arise as to why the data are missing arise—in particular, is the probability of a measurement vector being incomplete related to the unknown class (see Section 9.5)? A slightly more sophisticated approach is to try to substitute synthetic values for the missing items. At least then one has a complete data set to which one can apply standard classification methods. The danger here is that the substituted values may lead to distortion of the underlying distributions (this may not be a problem: if an incomplete new case is completed by the same process, it is as if it, too, has been sampled from the same distorted distribution). The risk of distortion can sometimes be sidestepped if a classification method based on a global model is being used. Then one can seek substitute values which are the same as those that would be predicted by the model, so they do not influence the model's parameters. The EM algorithm (Dempster *et al.*, 1977) illustrates this approach. A further risk with the use of substitute values is that one may subsequently underestimate the variance in the data.

Yet a third simple approach to missing values of measurements is to treat 'missing' as a legitimate response category. Indeed, it is often the case that the fact that a value is missing conveys information about the likely class membership. In one problem we studied, the ratio of the proportion missing in one class to the proportion missing in the other class varied from 0.28 to 2.33, depending on the variable chosen.

If components are missing from the measurement vector of a new point to be classified, unless substitute values are inserted, the classification must be made in a marginal subspace of the complete measurement space. This poses a serious problem for methods based on global models, such as neural networks, classical discriminant analysis and logistic regression. If only a few patterns of missing data are expected, one may construct appropriate rules for each relevant marginal subspace. Generally, however, this will not be possible. In such cases a model based on assuming independence of the predictor variables has the advantage that predictions in its marginal subspaces can be directly computed. For tree methods, proxy variables have been used, so that one is not stopped at an internal node of the tree. If 'missing' is regarded as a separate legitimate response category then no problem arises (though global models based on an assumed form for the population distributions may encounter problems). Nearest neighbour methods can readily be applied when the data are incomplete, though difficulties will arise if the design set data have been preprocessed (to permit accelerated search, for example).

Other aspects of the data which influence choice of rule are the number of classes, the extent of overlap (how hard is the problem), the relative sizes of the classes, the number of variables, the complexity of the decision surface and the relationship between this complexity and the number of variables. There is also a relationship between the complexity of the decision surface one can hope to fit and the size of the available design set; with a small sample it would be hoping for too much to expect to fit a decision surface with an abundance of twists and turns. Again we must remember that we want to model the underlying decision surface, not the design data themselves.

Large design sets permit one (in principle, at least) to identify small differences between classification rules. Indeed, large design sets are necessary to take advantage of the possibilities offered by a highly parametrised model (such as a neural network or nearest neighbour classifier), in terms of, for example, slight deviations of the decision surface from linearity. On the other hand, caution is required to avoid exaggerated claims. Hand and Henley (1996a) argue that small differences found during the design phase of a classifier (even if very highly statistically significant) may not be relevant to the practical application of the classifier, essentially because of phenomena such as population drift, subtle breakdown of assumptions when very large samples are involved, and so on. This means that there is a limit beyond which it is not worth pushing the development of classification rules: any apparent improvement in the decision surface is swamped by other kinds of inaccuracies. These sorts of issues are clearly very problem dependent, and need to be considered in context.

11.5   SEQUENTIAL METHODS

Sometimes it is useful to consider the variables sequentially. One possible reason for this is cost. In a medical situation, if measurement involves some complicated and expensive test procedure, one will naturally want to use as few measurements as possible. Time is another factor; if most objects can be accurately classified using only a few measurements, it is wasteful to collect other measurements on these objects. Instead one can include a 'doubtful' class and measure more variables only on those regarded as doubtful (the reject option). In principle, any classification method can be used in this sequential way. Indeed, one could even mix classifier types, using one approach for the initial classification, a different approach for the second classification, yet another for the third, and so on. This can be effective if speed matters—then one can use a very quick (and hence relatively inaccurate) method for the first pass. (And this principle of delaying the classification on a doubtful class can be applied whether or not the variables are taken in stages.)

If the same kind of classification rule is to be used at every stage, some methods are more suitable than others. Trees are, perhaps, the obvious class, since by definition they use their variables sequentially (except the more sophisticated versions involving model averaging). Methods such as discriminant analysis, logistic regression and neural networks are less straightforward to apply here, essentially requiring separate classification rules to be constructed. Nearest neighbour and kernel methods also effectively require separate rules (they need new metrics which include all the variables used at each stage).

Selection of the variables is just one way in which sequential notions can enter. A second is in dynamic updating of rules as time progresses. We have already referred to population drift—the tendency of populations to evolve with time. This is less crucial in some domains (e.g. medical diagnosis) than others (e.g. credit scoring) but, when it does occur, it can mean that the classification rule rapidly degrades. It would be nice if the rule could be updated as new cases achieve known true classifications. All methods which involve extensive preprocessing of the data face difficulties in this regard, and the only real approach is complete recalculation. But methods such as nearest neighbour (using, for example, the Euclidean metric) can be straightforwardly applied.

11.6   SPEED

Two speeds are relevant to classification problems: the speed with which a rule can be constructed and the speed with which a classification can be made. Again, those methods requiring the minimum amount of preprocessing will lead to quickest rule construction; two examples are nearest neighbour methods, if little effort is devoted to finding a good metric, and linear discriminant analysis. Methods such as neural networks, with many parameters to be estimated, can be painfully slow. Trees, also, can be slow because they require complex searches.

As far as classifying new objects goes, methods such as linear discriminant analysis, which are based on a simple linear combination of the variables, again win (providing little feature extraction is required). Neural networks can be designed to work in parallel, and can thus be very fast, but most implementations use serial simulations of parallel machines. Trees are also fast. Nearest neighbour methods, on the other hand, are relatively slow, again because of the searching and sorting involved. Methods exist (see Chapter 5) for accelerating such searches, either by preprocessing the data to remove redundant design set objects or by preprocessing them to yield a 'multivariate ordering' and using a branch and bound algorithm.

Perhaps it is worth adding that what is 'slow' and what is 'fast' depends very much on the application. A nearest neighbour method may be relatively slow, but if it can make a classification in a tenth of a second, it might be fast enough for most applications.

11.7 SEPARABILITY

The classes in a classification problem are described as *perfectly separable* or simply *separable* if the support regions of the population distributions do not intersect. This means that, at any given point of the measurement space, objects from only one class will be observed. Unlike the machine learning and computational learning theory communities, the statistical community has always assumed nonseparable classes; whereas the former have put a lot of effort into separable cases. The machine learning community, for example, has placed great emphasis on complete accuracy in design set classifications, something which the statistical community has assumed may not be possible. Again, which approach is more appropriate will depend on the problem. In 'natural' problems, where one cannot guarantee that the chosen variables span the space of differences between the objects, the statistical perspective seems more appropriate. In contrast, in 'artificial', 'synthetic' or 'logical' domains—problems created by humans, for example, in which there is a correct class for a given vector of describing variables—the machine learning perspective may be more natural. An example is given in Michie (1989) and cited in Venables and Ripley (1994, Chapter 13) involving whether a space-shuttle pilot should use the autolander or land manually.

Another illustration, for tree classifiers, lies in the difference between Breiman *et al* (1984) and Schuerman and Doster (1984). Breiman *et al*. implicitly assume that intermediate nodes need not perfectly classify any objects: the partitioning maximises separability in some sense, without necessarily leading to pure offspring nodes. In contrast, Schuerman and Doster assume that the norm for tree classifiers is to make certain 'correct' classifications (on the design set) at intermediate nodes. Thus they say (p. 359):

'The multistage recognition system, in fact, is an implementation of the powerful decision strategy of stepwise exclusion of alternatives. Among the potential results of the decision process, only those are followed up which cannot be positively excluded at the current state of the decision procedure. From node to node along the path through the decision tree, the number of alternatives is monotonically decreased, which results in a substantial simplification of the recognition task, until any ambiguity is removed at the terminal nodes.'

Schuerman and Doster (1984) then describe, as a radical departure from the machine learning/pattern recognition norm, a 'soft' classification approach, (p. 360): 'At every node, a vector of certain conditional *a posteriori* probabilities is generated, on which, in conventional tree classifier, a local decision would be based. These values are then transmitted to the subsequent descendant nodes.' They summarise (p. 367):

'The fundamental difference between tree classification as proposed here and the conventional tree classifier design lies in the replacement of the hard-decision strategy by a soft-decision one. At the output ports of the different nodes of the tree, *a posteriori* probabilities are provided. The conventional tree classifier incorporates a maximum selection and decision into every node. Thus, the only branch followed up is the most probable one in the present situation. The soft-decision concept admits alternatives in every partial decision.'

From a statistical perspective, see for example Sturt (1981), Mabbett *et al.* (1980), as well as Breiman *et al.* (1984), the 'soft-decision' strategy is the norm. We speculate that this has arisen because statisticians, as well as being concerned with building good classification rules, essentially see the problem in terms of the underlying probability distributions. Of course, even if the classes are perfectly separable, there still exists the problem of generalising to regions of the measurement space not containing design set points (see Chapter 1).

11.8   EMPIRICAL STUDIES OF CLASSIFICATION RULES

Supervised classification (like forecasting) is a natural area in which comparative empirical tests can be undertaken. A particular performance criterion is identified—error rate is overwhelmingly the most popular, with Brier score a poor second—and different methods are applied to a variety of data sets, both real and simulated, to see which does best. In view of this, it is hardly surprising to find that many such empirical studies have been carried out.

Some cautionary notes are in order concerning these studies. Firstly, they are often undertaken by researchers with a special interest and expertise in a particular method or class of methods. Leaving aside the possibility of deliberate bias (which would ultimately be pointless), there arises the possibility that the comparisons may be unfair because of that very expertise: researchers are likely to use the 'best' method, in some sense, amongst the class of methods on which they are expert, and less likely to use the corresponding 'best' from other classes of methods.

Secondly, especially in simulation studies, data sets are likely to be biased towards particular methods. For example, if all the data were generated from two multivariate normal classes with equal covariance matrices, one might expect linear discriminant analysis to do well. This problem generalises to real data sets; we have stressed again and again that the 'best' method depends on the problem. This means that different samples of problems could lead to different recommendations regarding which method is best. The most one should realistically hope for is some general recommendations matching up characteristics of the data and the problems with methods—along the lines of those presented in preceding sections in this chapter.

Thirdly, as noted in Hand and Henley (1996b):

'Classification techniques can be compared in two situations: (i) as used in standard "commercial" applications, by someone who is competent but not a leading researcher in the methodology being used, and (ii) as used by someone who is actively involved in developing the methodology and will take advantage of subtleties of which the commercial user may be unaware (perhaps even developing new features to fit the characteristics of the data). It would not be surprising if comparisons undertaken in the two different ways led to different results.'

Hand and Henley conclude that, if a commercial organisation is considering adopting some new classification technology, it should be evaluated in-house by the people who will use it.

A nice illustration of the difficulties that can arise in comparative studies is given by Michie *et al.* (1994, p. 65) when explaining how they selected the tree-based classification rules for their study, 'It should be emphasised that their selection was necessarily to a large extent arbitrary, having more to do with the practical logic of coordinating a complex and geographically distributed project than with judgements of merit or importance.' These sorts of points raise all sorts of questions about the practical relevance of the conclusions.

In an effort to overcome at least some of the difficulties outlined above, several studies have been set up in which different groups of researchers pit themselves (and their favourite methods) against others, using data sets provided by an independent source.

We will not attempt a global summary of comparative studies here. There have been many such studies (not counting those which develop new methods and include small empirical comparisons) so a global summary would merit an extensive review paper or book in its own right. Instead we will briefly indicate the flavour of some of the studies, referring the reader to the original accounts for details. But before doing so, it is worth restating our belief that sensitive and sophisticated use of just about any method, allied with a good understanding of the data, can lead to effective classification accuracy. Examples of empirical comparative studies are Gessaman and Gessaman (1972), Batchelor and Hand (1976), Titterington *et al.* (1981), Hand (1983), Mingers (1989), Atlas *et al.* (1990), Buntine and Niblett (1992), Brown *et al.* (1993), and Michie *et al.* (1994).

Titterington *et al.* (1981) compared independence models, models which included categorical interactions, latent class models, kernel methods, linear and quadratic discrimination, and logistic discrimination on data from 1000 patients with severe head injuries. The data included mixed types (categorical and continuous) and also suffered from missing values. Both the paper and the ensuing discussion are well worth reading, despite being over 15 years old. It is interesting to contrast the state of knowledge then and now. The independence model and linear discriminant analysis performed surprisingly well in comparison with the categorical interaction and kernel models: a phenomenon we would now explain in terms of the flexibility of the model relative to the complexity of the decision surface and the size of the design sets.

Michie *et al.* (1994) summarise the results of a large collaborative project (Statlog), involving six academic and six industrial laboratories, and funded by the European Community during the period 1990 to 1993. Problems of bias were avoided by a rigorous procedure for evaluating performance:

'First an agreed data format was established, algorithms were "deposited" at one site, with appropriate instructions; this version would be used in the case of any future dispute. Each dataset was then divided into a training set and a testing set, and any parameters in an algorithm would be "tuned" or estimated *only* by reference to the training set. Once a rule had been determined, it was then applied to the test data. This procedure was validated at another site by another (more naive) user for each dataset in the first phase of the Project. This ensured that the guidelines for parameter selection were not violated, and also gave some information on the ease-of-use for a non-expert in the domain.' Michie *et al.*, (1994, p. 4)

About 20 methods were applied to about 20 data sets. The methods included linear and quadratic discriminant analysis, logistic discriminant analysis, projection pursuit methods, kernel and nearest neighbour methods, causal network methods, tree methods, radial basis function methods and other neural network structures.

The studies of Titterington *et al.* (1981) and Michie *et al.* (1994) are unusual in that they compare a wide range of methods. Such broadly based methods clearly require extensive effort, either many collaborators or much time. More typical are studies which compare a restricted set of methods. Hand (1983), for example, compared linear discriminant analysis with kernel methods, and Brown *et al.* (1993) compared tree classifiers with neural networks. Hand (1983), applying the methods to multivariate binary data sets, found little difference in estimated true error rates. Brown *et al.* (1993) studied multimodal data sets and concluded: 'The results show that both methods produce comparable error rates but that direct application of either method will not necessarily produce the lowest error rate. In particular, we improve decision tree results with multi-variable splits and we improve backpropagation neural networks with feature selection and mode identification.' This illustrates the point made above, that superior performance can be achieved by someone who is expert in the relevant methodology.

In the empirical studies reported in Atlas *et al.* (1990), neural networks performed at least as well as the tree methods used, although they conclude there is insufficient evidence to provide strong support for either method alone. This seems hardly surprising in view of the theme of this book—different problems require different solutions.

Real data sets are attractive just because they are real. One knows that the data does not artificially favour some method and that the problems are genuine. On the other hand, real data sets have the disadvantage that the true properties of the underlying distributions are unknown so that, for example, true error rates have to be estimated (perhaps by resampling methods, such as leaving-one-out or boot-strap); this introduces extra inaccuracies into the conclusions. Simulated data sets do not have such problems, and many studies are therefore based on simulated data. One danger here, as mentioned above, is that simulated data sets may artificially match the model forms implicitly adopted by some class of methods, so the conclusions are unfairly biased. Proper experimental design is also impor-tant in simulation studies. The factors to be included need careful consideration, along with their ranges and the number of levels for each. These should be considered in the context of the sorts of problems one is interested in solving. It is unrealistic to hope for conclusions which are valid in all practical situations. A rare example of a well-designed simulation study is that of Ringrose and Krza-nowski (1991).

11.9 STATISTICS VERSUS MACHINE LEARNING

As we have remarked elsewhere, one of the exciting aspects of work on super-vised classification is that it has benefited from the efforts of several different intellectual communities. Researchers in statistics and computer science, espe-cially those in machine learning, pattern recognition and computational learning theory, have all taken the construction of classification rules from a design sample as a fundamental question.

Several differences in emphasis can be identified in the work of these disciplines:

• The computer science and pattern recognition community has emphasised *adaptive* estimation techniques, even going so far as to use the term *learning* for the estimation of parameters (sometimes calling the design set a learning or training set). This meant that they could handle very large data sets, since only the current parameter estimates and the latest data point needed to be kept in memory at one time.
• Whereas the earlier statistical work emphasised continuous measurement variables, the pattern recognition work has always recognised the importance of discrete (especially binary) variables.
• Early work in the computer science community set great store by *separability* of classes. Early convergence proofs for the estimates of parameters proved

that the system would converge to separating surfaces when these existed. Statisticians, however, were seldom interested in such a property, believing that separability rarely, if ever, occurred in practice.

- Statisticians always saw the problem as one of inference, as one of generalising from the design set to the larger population from which this set was drawn. In particular, this meant that the classification partition was seen as the result of imposing a threshold on estimated probabilities of class memberships. In contrast, the early pattern recognition community focused on the design set itself, and put its effort into finding a rule which separated the classes in this set. Inevitably this led to overfitting, a problem which is much more easily explained in terms of the inferential model. Interestingly enough, the problems associated with overfitting recurred a decade or two later with the resurgence of interest in multilayer neural networks.
- Compared with the statistical community, the pattern recognition community placed much greater emphasis on the algorithms, perhaps partly because they had their home in computers. Anyone concerned with adaptive estimation will inevitably be concerned with algorithms.
- The computer science work (or, at least, the artificial intelligence work) on classification trees has put greater emphasis on interpretability than the statistical work, which has been more concerned with classification accuracy. This might be because the tree construction was often a mere knowledge acquisition step on the way to constructing an expert system, where interpretability is a defining characteristic.
- The statistical work has always placed great store in mathematical rigour (convergence proofs, optimality properties, asymptotic results, and so on) whereas the computer science work has had a more ad hoc quality—try it and see if it works in practice.

11.10   CONCLUSION

Throughout this book we have attempted to illustrate that different problems and different application domains have different properties, requiring different solutions. The range of such differences is unlimited, and in this book we could only touch on some of the more important. Examples of other situations, leading to unusual and challenging problems in the construction of classification rules, are how to cope with errors in the 'true' classes, how to use unclassified data to improve predictive accuracy, applications involving very many classes, perhaps each with few design set points, confidence intervals on error rates and other performance measures, and predictive discrimination, in which inaccuracy of parameter estimates is included in the predictions. Some of these topics are dealt with by the books described in Chapter 1 and the references at the end of the other chapters.

The opening sentence of this book posed the question, Which is the best type of classification rule? We can now answer that question. The answer is: It depends ...

References

Aczel J. and Daroczy Z. (1975) *On measures of information and their characterization*. New York: Academic Press.

Aeberhard S., Coomans D., and de Vel O. (1993) Improvements to the classification performance of RDA. *Journal of chemometrics*, **7**, 99–115.

Aha D.W., Kibler D., and Albert M.K. (1991) Instance-based learning algorithms. *Machine Learning*. **6**, 37–66.

Aitchison J. and Aitkin C.G.G. (1976) Multivariate binary discrimination by the kernel method. *Biometrika*, **63**, 413–420.

Aitchison J., Habbema J.D.F., and Kay J.W. (1977) A critical comparison of two methods of statistical discrimination. *Applied Statistics*, **26**, 15–25.

Akaike H. (1973) Information theory and an extension of the maximum likelihood principle. In *Second International Symposium on Information Theory*, ed. B.N. Petrov and F.Cáski. Budapest: Akademiai Kaidó. Reprinted in *Breakthroughs in Statistics*, ed. S.Kotz and N.L. Johnson (1992) Vol. I, New York: Springer, pp. 599–624.

Ali F. and Pavlidis T. (1977) Syntactic recognition of handwritten numerals. *IEEE Transactions on Systems, Man, and Cybernetics*, **7**, 537–541.

Anderberg M.R. (1973) *Cluster Analysis for Applications*. New York: Academic Press.

Anderson J.A. (1969) Constrained discrimination between *k* populations. *Journal of the Royal Statistical Society, Series B*, **31**, 123–139.

Anthony M. and Biggs N. (1992) *Computational Learning Theory*. Cambridge: Cambridge University Press.

Atlas L., Cole R., Muthusamy Y., Lippman A., Connor J., Park D., El-Sharkawi M., and Marks R.J. (1990) A performance comparison of trained multilayer perceptrons and trained classification trees. *Proceedings of the IEEE*, **78**, 1614–1619.

Bartlett M.S. (1947) Multivariate analysis. *Journal of the Royal Statistical Society, Supplement* **9**, 176–190.

Bartlett M.S. (1951) The goodness of fit of a single hypothetical discriminant function in the case of several groups. *Annals of Eugenics*, **16**, 199–214.

Batchelor B.G. (1968) *Learning Machines for Pattern Recognition*. PhD thesis. University of Southampton.

Batchelor B.G. and Hand D.J. (1976) A pattern recognition competition, *Proceedings of the Third International Joint Conference on Pattern Recognition*, San Diego.

Berkson J. (1944) Application of the logistic function to bio-assay. *Journal of the American Statistical Association*, **39**, 357–365.

Bernardo J.M. and Smith A.F.M. (1994) *Bayesian Theory*. Chichester: John Wiley.

Bhattacharyya A. (1943) On a measure of divergence between two statistical populations defined by their probability distributions. *Bulletin of the Calcutta Mathematical Society*, **35**, 99–109.

Bishop C.M. (1995) *Neural Networks for Pattern Recognition.* Oxford: Clarendon Press.

Bishop Y.M.M., Fienberg S.E., and Holland P.W. (1975) *Discrete Multivariate Analysis: Theory and Practice.* Cambridge, MA: MIT Press.

Breiman L., Freidman J.H., Olshen R.A., and Stone C.J. (1984) *Classification and Regression Trees.* Belmont, CA: Wadsworth.

Brodley C.E. and Smyth P. (1996) Applying classification algorithms in practice. *Statistics and Computing,* forthcoming.

Brown D.E., Corruble V., and Pittard C.L. (1993) A comparison of decision tree classifiers with backpropagation neural networks for multimodal classification problems. *Pattern Recognition,* **26**, 953–961.

Bryson A.E. and Ho Y.-C. (1969) *Applied Optimal Control.* New York: Blaisdell.

Buntine W.L. (1992) Learning in classification trees. *Statistics and Computing,* **2**, 63–73.

Buntine W.L. and Niblett T. (1992) A further comparison of splitting rules for decision-tree induction. *Machine Learning,* **8**, 75–85.

Burgess E.W. (1928) Factors determining success or failure on parole. In *The Workings of the Intermediate-sentence Law and the Parole System in Illinois,* ed. A.A.Bruce, A.J.Harno, E.W.Burgess, and J.Landesco. Springfield: Illinois State Board of Parole.

Cacoullos T. (1966) Estimation of a multivariate density. *Annals of the Institute of Statistical Mathematics,* **18**, 179–189.

Campbell N.A. (1980) Shrunken estimators in discriminant and canonical variate analysis. *Applied Statistics,* **29**, 5–14.

Carothers A. and Piper J. (1994) Computer-aided classification of humna chromosomes: a review. *Statistics and Computing,* **4**, 161–171.

Chang P.C. and Afifi A.A. (1974) Classification based on dichotomous and continuous variables. *Journal of the American Statistical Association,* **69**, 336–339.

Cheeseman P. and Oldford R.W. (eds.) (1994) *Selecting Models from Data.* New York: Springer-Verlag.

Chernoff H. (1952) A measure of asymptotic efficiency for tests of a hypothesis based on a sum of observations. *Annals of Mathematical Statistics,* **23**, 493–507.

Chidananda Gowda K. and Krishna G. (1979) The condensed nearest neighbour rule using the concept of mutual nearest neighbourhood. *IEEE Transactions on Information Theory,* **25**, 488–490.

Chiogna M. (1994) Probabilistic symbolic classifiers: an empirical comparison from a statistical perspective. In *ECML-94 Seventh European Conference on Machine Learning,* Catania, Italy, 5–8 April, 1994, ed. G.Nakhaeizadeh and C.Taylor.

Choi S.C., Muizelaar J.P., Barnes T.Y., Marmarou A., Brooks D.M., and Young H.F. (1991) Prediction tree for severely head-injured patients. *Journal of Neurosurgery,* **75**, 251–255.

Chou P.A. (1988) *Applications of Information Theory to Pattern Recognition and the Design of Decision Trees and Trellises.* PhD thesis. Stanford University.

Chou P.A. (1991) Optimal partitioning for classification and regression trees. *IEEE Transactions on Pattern Analysis and Machine Intelligence,* **13**, 340–354.

Clark L.A. and Pregibon D. (1992) Tree-based models. In *Statistical Models in S,* ed J.M.Chambers and T.J.Hastie. Belmont, CA: Wadsworth, pp. 377–420.

Clarke P. and Niblett T. (1989) The CN2 induction algorithm. *Machine Learning,* **3**, 261–283.

Cohen E., Hull J.J., and Srihari S.N. (1990) Understanding handwritten text in a structured environment: determining ZIP codes from addresses. *International Journal of Pattern Recognition and Artificial Intelligence,* **5**, 221–264.

Collett D. (1991) *Modelling Binary Data*. London: Chapman and Hall.

Coomans D. and Broeckaert I. (1986) *Potential Pattern Recognition*. Letchworth: Research Studies Press.

Cooper C., Shah S., Hand D.J., Adams J., Compston J., Davie M., and Woolf A. (1991) Screening for vertebral osteoporosis using individual risk factors. *Osteoporosis*, **2**, 48–53.

Copas J.B. and Li H.G. (1997) Inference for non-random samples. *Journal of the Royal Statistical Society, Series B*, **59**, in press.

Cover T.M. (1965) Geometrical and statistical properties of systems of linear inequalities with applications in pattern recognition. *IEEE Transactions on Electronic Computers*, **14**, 326–334.

Cover T.M. and Hart P.E. (1967) Nearest neighbour pattern classification. *IEEE Transactions on Information Theory*, **13**, 21–27.

Crowder M. J. and Hand D. J.(1990) *Analysis of Repeated Measures*. London: Chapman and Hall.

Davis K.H., Biddulph R., and Balshek S. (1952) Automatic recognition of spoken digits. *Journal of the Acoustical Society of America*, **24**, 637–642.

Davis R.H., Edelman D.B., and Gammerman A.J. (1992) Machine-learning algorithms for credit-card applications. *IMA Journal of Mathematics Applied in Business and Industry*, **4**, 43– 51.

Dawes R.M. and Corrigan B. (1974) Linear models in decision making. *Psychological Bulletin*, **81**, 95–106.

Dawid A.P. (1976) Properties of diagnostic data distributions. *Biometrics*, **32**, 647–658.

Dawid A.P. (1982) The well-calibrated Bayesian. *Journal of the American Statistical Association*, **77**, 605–610.

DeGroot M. and Fienberg S.E. (1983) The comparison and evaluation of forecasters. *The Statistician*, **32**, 12–22.

DeLong E.R., DeLong D., and Clarke-Pearson D.L. (1988) Comparing the areas under two or more correlated receiver operating characteristic curves: a nonparametric approach. *Biometrics*, **44**, 837–845.

Dempster A.P. (1972) Covariance selection. *Biometrics*, **28**, 157–175.

Dempster A.P., Laird N.M., and Rubin D. (1977) Maximum likelihood from incomplete data via the EM algorithm. *Journal of the Royal Statistical Society*, **39**, 1–38.

Devijver P.A. and Kittler J. (1982) *Pattern Recognition: a Statistical Approach*. Englewood Cliffs, N J: Prentice Hall.

Draper D. (1995) Assessment and propagation of model uncertainty (with discussion). *Journal of the Royal Statistical Society, Series B*, **57**, 45–97.

Duda R. and Hart P. (1973) *Pattern Classification and Scene Analysis*. New York: John Wiley.

Dunn G. (1989) *Design and Analysis of Reliability Studies*. New York: Oxford University Press.

Dunn G., Sham P.C., and Hand D.J. (1993) Statistics and the nature of depression, *Journal of the Royal Statistical Society, Series A*, **156**, 63–87.

Durand D. (1941) *Risk elements in consumer instalment financing*. New York: National Bureau of Economic Research.

Efron B. (1975) The efficiency of logistic regression compared to normal discriminant analysis. *Journal of the American Statistical Association*, **70**, 892–898.

Efron B. (1982) *The Jackknife, the Bootstrap, and other Resampling Plans*. Philadelphia: SIAM.

Efron B. (1983) Estimating the error rate of a prediction rule: improvement on cross-validation. *Journal of the American Statistical Association*, **78**, 316–330.

Efron B. and Gong G. (1983) A leisurely look at the bootstrap, the jackknife, and cross-validation. *American Statistician*, **37**, 36–48.

Efron B. and Tibshirani J. (1993) *An Introduction to the Bootstrap*. London: Chapman and Hall.

Epanechnikov V.A. (1969) Non-parametric estimation of a multivariate probability density. *Theory of Probability and its Applications*, **14**, 153–158.

Everitt B.S. (1974) *Cluster Analysis*. London: Heinemann.

Fahlman S.E. (1989) Fast-learning variations on back-propagation: an empirical study. In *Proceedings of the 1988 Connectionist Models Summer School*, Pittsburgh, 1988, ed. D.Touretzky, G. Hinton, and T.Sejnowski. San Mateo: Morgan Kaufmann, pp. 38–51.

Ferguson J.D. (1980) Hidden Markov analysis: an introduction. In *Hidden Markov Models for Speech*. Princeton, NJ: Institute for Defense Analyses.

Fielding A. (1977) Binary segmentation: the automatic interaction detector and related techniques for exploring data structure. In *The analysis of survey data, Vol.1: Exploring data structures*, C.A. O'Muircheartaigh and C.Payne. New York: John Wiley.

Fisher R.A. (1936) The use of multiple measurements in taxonomic problems. *Annals of Eugenics*, **7**, 179–188.

Fitzmaurice G.M. and Hand D.J. (1987) A comparison of two average conditional error rate estimators. *Pattern Recognition Letters*, **6**, 221–224.

Fitzmaurice G.M., Krzanowski W.J., and Hand D.J. (1991) A Monte Carlo study of the 632 bootstrap estimator of error rate. *Journal of Classification*, **8**, 239–250.

Fix E. and Hodges J.L. (1951) Discriminatory analysis—nonparametric discrimination: consistency properties. *Report No. 4, Project no. 21–29–004*, USAF School of Aviation Medicine, Randolph Field, Texas. Reprinted in *International Statistical Review*, **57** (1989) 238–247.

Flury B.D. (1988) *Common Principal Components and Related Multivariate Models*. New York: John Wiley.

Flury B.D. (1995) Developments in principal component analysis. In *Recent advances in descriptive multivariate analysis*, ed. W.J.Krzanowski. Oxford: Clarendon Press.

Forgie J.W. and Forgie C.D. (1959) Results obtained from a vowel recognition computer program. *Journal of the Acoustical Society of America*, **31**, 1480–1489.

Frank I.E. (1987) Intermediate least squares regression method. *Chemometrics and Intelligent Laboratory Systems*, **1**, 233–242.

Frank I.E. and Friedman J.H. (1989) Classification: oldtimers and newcomers. *Journal of Chemometrics*, **3**, 463–475.

Friedman J.H. (1989) Regularized discriminant analysis. *Journal of the American Statistical Association*, **84**, 165–175.

Friedman J.H. (1991) Multivariate adaptive regression splines (with discussion). *Annals of Statistics*, **19**, 1–141.

Friedman J.H. and Stuetzle W. (1981) Projection pursuit regression. *Journal of the American Statistical Association*, **76**, 817–823.

Fukunaga K. (1990) *Introduction to Statistical Pattern Recognition*, 2nd edn. San Diego: Academic Press.

Fukunaga K. and Flick T.E. (1984) An optimal global nearest neighbour metric. *IEEE Transactions on Pattern Recognition and Machine Intelligence*, **6**, 314–318.

Fukunaga K. and Kessell D.L. (1973) Nonparametric Bayes error estimation using unclassified samples. *IEEE Transactions on Information Theory*, **19**, 434–440.

Fukunaga K. and Narendra P.M. (1975) A branch and bound algorithm for computing *k*-nearest neighbours. *IEEE Transactions on Computers*, **24**, 750–753.

Gamba A.L., Gamberini G., Palmieri G., and Sanna R. (1961) Further experiments with PAPA. *Nuovo Cimento Supplement*, **20**, 221–231.

Ganesalingam S. and McLachlan G.J. (1980) Error rate estimation on the basis of posterior probabilities. *Pattern Recognition*, **12**, 405–413.

Garner D.M. and Garfinkel P.E. (1980) Socio-cultural factors in the development of anorexia nervosa. *Psychological Medicine*, **10**, 647–656.

Garnett J.M. and Yau S.S. (1977) Nonparametric estimation of the Bayes error of feature extractors using ordered nearest neighbour sets. *IEEE Transactions on Computers*. **26**, 46–54.

Gates G.W. (1972) The reduced nearest neighbour rule. *IEEE Transactions on Information Theory*, **18**, 431.

Gelfand S.B., Ravishankar C.S., and Delp E.J. (1991) An iterative growing and pruning algorithm for classification tree design. *IEEE Transactions on Pattern Analysis and Machine Intelligence*, **13**, 163–174.

Genest C. and McConway K.J. (1990) Allocating the weights in the linear opinion pool. *Journal of Forecasting*, **9**, 53–73.

Gersho A. and Gray R.M. (1992) *Vector Quantization and Signal Compression*. Dordrecht: Kluwer Academic.

Gessaman M.P. and Gessaman P.H. (1972) A comparison of some multivariate discrimination procedures. *Journal of the American Statistical Association*, **67**, 468–472.

Gilbert E.S. (1968) On discrimination using qualitative variables. *Journal of the American Statistical Association*, **63**, 1399–1412.

Glick N. (1978) Additive estimators for probabilities of correct classification. *Pattern Recognition*, **10**, 211–222.

Goldman L., Weinberg M., Weisberg M., Olshen R., Cook E.F., Sargent R.K., Lamas G.A., Dennis C., Wilson C., Deckelbaum L., Fineberg H., Stiratelli R., and the Medical House Staff at Yale–New Haven and Brigham Women's Hospital (1982) A computer derived protocol to aid in the diagnosis of emergency room patients with chest pain. *New England Journal of Medicine*, **307**, 588–596.

Gonzalez R.C. and Thomason M.G. (1978) *Syntactic Pattern Recognition: An Introduction*. London: Addison-Wesley.

Gordon A.D. (1981) *Classification*. London: Chapman and Hall.

Granum E. (1982) Application of statistical and syntactical methods of analysis and classification to chromosome data. In *NATO ASI Series: C81 Pattern Recognition Theory and Applications*, ed. J.Kittler. Dordrecht : D.Reidel pp. 373–398.

Granum E. and Thomason M.G. (1990) Automatically inferred Markov network models for classification of chromosomal band pattern structures. *Cytometry*, **11**, 26–39.

Granum E., Gerdes T., and Lundsteen C. (1981) Simple weighted density distributions, WDDs, for discrimination between G-banded chromosomes. *Proceedings of the Fourth European Chromosome Analysis Workshop*, Edinburgh.

Green B.F. (1977) Parameter sensitivity in multivariate methods. *Journal of Multivariate Behavioural Research*, **12**, 263–287.

Greenwood R.W. (1982) *Selective Incapacitation*. Santa Monica: Rand Corporation.

Habbema J.D.F., Hermans J., and van der Burgt A.T. (1974) Cases of doubt in allocation problems. *Biometrika*, **61**, 313–324.

Habbema J.D.F., Hilden J. and Bjerregaard B. (1978) The measurement of performance in probabilistic diagnosis I. The problem, descriptive tools, and measures based on classification matrices. *Methods of Information in Medicine*, **17**, 217–226.

Hand D.J. (1981a) *Discrimination and Classification*. Chichester: John Wiley.

Hand D.J. (1981b) Branch and bound in statistical data analysis. *Statistician*, **30**, 1–13.

Hand D.J. (1982) *Kernel Discriminant Analysis*. Letchworth: Research Studies Press.

Hand D.J. (1983) A comparison of two methods of discriminant analysis applied to binary data. *Biometrics*, **39**, 683–694.

Hand D.J. (1985) *Artificial Intelligence and Psychiatry*. Cambridge: Cambridge University Press.

Hand D.J. (1986) Recent advances in error rate estimation. *Pattern Recognition Letters*, **4**, 335–346.

Hand D.J. (1987a) A shrunken leaving-one-out estimator of error rate. *Computers and Mathematics with Applications*, **14**, 161–167.

Hand D.J. (1987b) Screening vs. prevalence estimation. *Applied Statistics*, **36**, 1–7.

Hand D.J. (ed.) (1990) *Annals of Mathematics and Artificial Intelligence*, **2**, special issue on artificial intelligence and statistics.

Hand D.J. (1993) (ed.) *Artificial Intelligence Frontiers in Statistics*. London: Chapman and Hall.

Hand D.J. (1994a) Deconstructing statistical questions (with discussion). *Journal of the Royal Statistical Society, Series A*, **157**, 317–356.

Hand D.J. (1994b) Recent results in pattern recognition theory and applications. *Current Topics in Pattern Recognition Research*, **1**, 113–123.

Hand D.J. (1994c) Assessing classification rules. *Journal of Applied Statistics*, **21**, 3–16.

Hand D.J. (1995a) Comparing allocation rules. *Statistics in Transition*, **2**, 137–150.

Hand D.J. (1995b) Screening for stratification in two-phase ('two-stage') epidemiological surveys. *Statistical Methods in Medical Research*, **4**, 263.

Hand D.J. (1996) Classification and computers: shifting the focus. *COMPSTAT 96, Proceedings in Computational Statistics*, ed. A. Prat, Physica-Verlag, 77–88.

Hand D.J. and Batchelor B.G. (1978) An edited condensed nearest neighbour rule. *Information Sciences*, **14**, 171–180.

Hand D.J. and Crowder M.J. (1996) *Practical Longitudinal Data Analysis*. London: Chapman and Hall.

Hand D.J. and Henley W.E. (1993) Can reject inference ever work? *IMA Journal of Mathematics Applied in Business and Industry*, **5**, 45–55.

Hand D.J. and Henley W.E. (1994) Inference about rejected cases in discriminant analysis. In *New Approaches in Classification and Data Analysis*, ed. E. Diday, Y. Lechevallier, M. Schader, P. Bertrand and B. Buntschy Berlin: Springer-Verlag, pp. 292–299.

Hand D.J. and Henley W.E. (1996a) Statistical classification methods in consumer credit scoring: a review *Journal of the Royal Statistical Society, Series A*, (In press).

Hand D.J. and Henley W.E. (1996b) Some developments in statistical credit scoring. In *Machine learning and statistics: the interface*, ed. G.Nakhaeizadeh and C.C.Taylor. New York: Wiley, forthcoming.

Hand D.J. and Taylor C.C. (1987) *Multivariate Analysis of Variance and Repeated Measures*. London: Chapman and Hall.

Hand D.J., Oliver J.J., and Lunn A.D. (1996) Discriminant analysis when the classes arise from a continuum. (Unpublished manuscript)

Hanley J.A. and McNeil B.J. (1982) The meaning and use of the area under a receiver operating characteristic (ROC) curve. *Radiology*, **143**, 29–36.

Härdle W. (1991) *Smoothing Techniques with Implementation in S*. New York: Springer Verlag.

Hart P.E. (1968) The condensed nearest neighbour rule. *IEEE Transactions on Information Theory*, **14**, 515–516.

Hasselbald V. and Hedges L.V. (1995) Meta-analysis of screening and diagnostic tests. *Psychological Bulletin*, **117**, 167–178.

Hastie T. and Pregibon D. (1990) *Shrinking trees*. Technical report. AT&T Bell Laboratories, Murray Hill, NJ 07974, USA.

Hastie T.J. and Tibshirani R.J. (1986) Generalized additive models (with discussion). *Statistical Science*, **1**, 297–318.

Hastie T.J. and Tibshirani R.J. (1990) *Generalized Additive Models*. London: Chapman and Hall.

Heitjan D.F. (1989) Inference from grouped continuous data: a review. *Statistical Science*, **4**, 164–183.

Heitjan D.F. and Rubin D.B. (1991) Ignorability and coarse data. *Annals of Statistics*, **19**, 2244–2253.

Helland I.S. (1988) On the structure of partial least squares regression. *Communications in Statistics*, **17**, 581–607.

Henery R.J. (1996) Combining classification rules. In *Machine learning and statistics : the interface*, ed. G.Nakhaeizadeh and C.Taylor. New York: John Wiley.

Henley W.E. (1995) *Statistical aspects of credit scoring*. PhD dissertation. The Open University, Milton Keynes, UK.

Henley W.E. and Hand D.J. (1996) A k-nearest neighbour classifier for assessing consumer credit risk. *Statistician*, **44**, 77–95.

Hertz J., Krogh A., and Palmer R.G. (1991) *Introduction to the Theory of Neural Computation*. Redwood City, CA: Addison-Wesley.

Hilden J., Habbema J.D.F., and Bjerregaard B. (1978a) The measurement of performance in probabilistic diagnosis II. Trustworthiness of the exact values of the diagnostic probabilities. *Methods of Information in Medicine*, **17**, 227–237.

Hilden J., Habbema J.D.F., and Bjerregaard B. (1978b) The measurement of performance in probabilistic diagnosis III. Methods based on continuous functions of the diagnostic probabilities. *Methods of Information in Medicine*, **17**, 238–246.

Hildritch C.J. and Rutovitz D. (1972) Normalisation of chromosome measurements. *Computers in Biology and Medicine*, **2**, 167–179.

Hilgers R.A. (1991) Distribution-free confidence bounds for ROC curves. *Methods of Information in Medicine*, **30**, 96–101.

Ho T.K., Hull J.J., and Srihari S.N. (1994) Decision combination in multiple classifier systems. *IEEE Transactions on Pattern Analysis and Machine Intelligence*, **16**, 66–75.

Ho Y.-C. and Kashyap R.L. (1965) An algorithm for linear inequalities and its applications. *IEEE Transactions on Electronic Computers*, **14**, 683–688.

Ho Y.-C. and Kashyap R.L. (1966) A class of iterative procedures for linear inequalities. *Journal of SIAM Control*, **4**, 112–115.

Huang Y.S. and Suen C.Y. (1995) A method of combining multiple experts for the recognition of unconstrained handwritten numerals. *IEEE Transactions on Pattern Analysis and Machine Intelligence*, **17**, 90–94.

Hunt E.B., Marin J., and Stone P. (1966) *Experiments in Induction*. New York: Academic Press.

Jacobs R.A. (1988) Increased rates of convergence through learning rate adaptation. *Neural Networks*, **1**, 295–307.

Jacobs R.A., Jordan M.I., Nowlan S.J., and Hinton G.E. (1991) Adaptive mixtures of local experts. *Neural Computation*, **3**, 79–97.

Jeffreys H. (1946) An invariant form for the prior probability in estimation problems. *Proceedings of the Royal Society A*, **186**, 453–461.

Joshi A.K. (1964) A note on a certain theorem stated by Kullback. *IEEE Transactions on Information Theory*, **10**, 93–94.

Kass G.V. (1980) An exploratory technique for investigating quantities of categorical data. *Applied Statistics*, **29**, 119–127.

Kittler J. and Devijver P.A. (1982) Statistical properties of error rate estimators in performance assessment of recognition systems. *IEEE Transactions on Pattern Analysis and Machine Intelligence*, **4**, 215–220.

Kowalski B. and Wold S. (1982) Pattern recognition in chemistry. In *Handbook of Statistics*, Vol.2, ed. P.R.Krishnaiah and L.Kanal. Amsterdam: North-Holland, pp. 673–697.

Krzanowski W.J. (1975) Discrimination and classification using both binary and continuous variables. *Journal of the American Statistical Association*, **70**, 782–790.

Krzanowski W.J. (1977) The performance of Fisher's linear discriminant function under non-optimal conditions. *Technometrics*, **19**, 191–199.

Krzanowski W.J. (1983a) Distance between populations using mixed continuous and categorical variables. *Biometrika*, **70**, 235–243.

Krzanowski W.J. (1983b) Stepwise location model choice in mixed-variable discrimination. *Applied Statistics*, **32**, 260–266.

Krzanowski W.J. (1984) On the null distribution of distance between two groups, using mixed continuous and categorical variables. *Journal of Classification*, **1**, 243–253.

Krzanowski W.J. (1986) Multiple discriminant analysis in the presence of mixed continuous and categorical data. *Computers and Mathematics with Applications*, **4**, 73–84.

Krzanowski W.J. (1987) A comparison between two distance-based discriminant principles. *Journal of Classification*, **4**, 73–84.

Krzanowski W.J. (1989) On confidence regions in canonical variate analysis. *Biometrika*, **76**, 107–116.

Krzanowski W.J. and Hand D.J. (1996) Assessing error rate estimators: the leave-one-out method reconsidered. *Australian Journal of Statistics*, submitted.

Krzanowski W.J. and Marriott F.H.C. (1995) *Multivariate Analysis, Part 2: Classification, Covariance Structures, and Repeated Measurements.*. London: Edward Arnold.

Krzanowski W.J., Jonathan P., McCarthy W.V., and Thomas M.R. (1995) Discriminant analysis with singular covariance matrices: methods and applications to spectroscopic data. *Applied Statistics*, **44**, 101–115.

Kullback S. (1959) *Information Theory and Statistics*. New York: John Wiley.

Lachenbruch P.A. (1965) *Estimation of error rates in discriminant analysis*. PhD thesis, University of Los Angeles.

Lachenbruch P.A. (1975) *Discriminant analysis*. New York: Hafner Press.

Landeweerd G., Timmers T., Gersema E., Bins M., and Halic M. (1983) Binary tree versus single-level tree classification of white blood cells. *Pattern Recognition*, **16**, 571–577.

Laughlin J.E. (1978) Comment on 'Estimating coefficients in linear models: it don't make no nevermind'. *Psychological Bulletin*, **85**, 247–253.

Li X. and Dubes R.C. (1986) Tree classifier design with a permutation statistic. *Pattern Recognition*, **19**, 229–235.

Lissack T. and Fu K.S. (1976) Error estimation in pattern recognition. *IEEE Transactions on Information Theory*, **22**, 34–35.

Little R.J.A. and Rubin D.B. (1987) *Statistical Analysis with Missing Data*. New York: John Wiley.

Loh W.-Y. and Vanichesetakul N. (1988) Tree stuctured classification via generalized discriminant analysis. *Journal of the American Statistical Association*, **83**, 715–725.

Longford N. (1993) *Random Coefficient Models*. Oxford: Clarendon Press.

Lundsteen C., Gerdes T., Granum E., Philip J., and Philip K. (1981) Automatic chromosome analysis; karyotyping of banded human chromosomes. *Clinical Genetics*, 19, 26–36.

Mabbett A., Stone M., and Washbrook J. (1980) Cross-validatory selection of binary variables in differential diagnosis. *Applied Statistics*, 29, 198–204.

McCullagh P. and Nelder J.A. (1989) *Generalized Linear Models*. London: Chapman and Hall.

McLachlan G.J. (1977) A note on the choice of a weighting function to give an efficient method for estimating the probability of misclassification. *Pattern Recognition*, 9, 147–149.

McLachlan G.J. (1987) Error rate estimation in discriminant analysis: recent advances. In *Advances in Multivariate Statistical Analysis*, ed. A.K.Gupta. Dordrecht: D. Reidel, pp. 233–252.

McLachlan G.J. (1992) *Discriminant Analysis and Statistical Pattern Recognition*. New York: John Wiley.

Mahoney J.J. and Mooney R.J. (1991) Initializing ID5R with a domain theory: some negative results. Report 91-154. Department of Computer Science, University of Texas at Austin.

Mallows C. (1973) Some comments on C_p. *Technometrics*, 15, 661–675.

Mandler E. and Schurman J. (1988) Combining the classification results of independent classifiers based on the Dempster-Shafer theory of evidence. In *Pattern Recognition and Artificial Intelligence*, ed. E.Gelsema and L.Kanal. New York: Elsevier, pp. 381–393.

Marron J.S. and Nolan D. (1989) Canonical kernels for density estimation. *Statistics and Probability Letters*, 7, 195–199.

Matusita K. (1955) Decision rules based on the distance for problems of fit, two samples and estimation. *Annals of Mathematical Statistics*, 26, 631–640.

Metz C.E., Wang P.-L., and Kronman H.B. (1984) A new approach for testing the significance of differences between ROC curves measured from correlated data. In *Information Processing in Medical Imaging*, Vol. VIII, ed. F. Deconick. The Hague: Martinus Nijhof, pp. 432–445.

Michalski R.S. and Chilauski R.L. (1980) Knowledge acquisition by encoding expert rules versus computer induction from examples: a case study involving soybean pathology. *International Journal of Man–Machine Studies*, 12, 63–87.

Michie D. (1989) Problems of computer-aided concept formation. In *Applications of Expert Systems*, 2, ed. J.R.Quinlan. Glasgow:Turing Institute Press/Addison-Wesley, pp. 310–333.

Michie D., Spiegelhalter D.J., and Taylor C.C. (1994) *Machine Learning, Neural and Statistical Classification*. New York: Ellis Horwood.

Miller R.G. (1962) Statistical prediction by discriminant analysis. *Meteorological Monographs*, 4, No. 25.

Mingers J. (1989) An empirical comparison of selection measures for decision-tree induction. *Machine Learning*, 3, 319–342.

Minsky M. and Papert S. (1969) *Perceptrons: An Introduction to Computational Geometry*. Cambridge, MA: MIT Press.

Moore D.H. (1973) Evaluation of five discrimination procedures for binary variables. *Journal of the American Statistical Association*, 68, 399–404.

Morgan J.N. and Messenger R.C. (1973) *THAID: A Sequential Search Programme for the Analysis of Nominal Scale Dependent Variables*. Institute for Social Research, University of Michigan, Ann Arbor.

Morgan J.N. and Sonquist J.A. (1963) Problems in the analysis of survey data, and a proposal. *Journal of the American Statistical Association*, **58**, 415–434.

Mori S., Suen C.Y., and Yamamoto K. (1992) Historical review of OCR research and development. *Proceedings of the IEEE*, **80**, 1029–1058.

Murphy A.H. (1972a) Scalar and vector partitions of the probability score: Part I, Two-state situation. *Journal of Applied Meteorology*, **11**, 273–282.

Murphy A.H. (1972b) Scalar and vector partitions of the probability score: Part II, *N*-state situation. *Journal of Applied Meteorology*, **11**, 1183–1192.

Murphy A.H. (1973) A new vector partition of the probability score. *Journal of Applied Meteorology*, **12**, 595–600.

Myles J.P. and Hand D.J. (1990) The multi-class metric problem in nearest neighbour discrimination rules. *Pattern Recognition*, **23**, 1291–1297.

Nagy G. (1992) At the frontiers of OCR. *Proceedings of the IEEE*, **80**, 1093–1100.

Narendra P.M. and Fukunaga K. (1977) A branch and bound algorithm for feature subset selection. *IEEE Transactions on Computers*, **26**, 917–922.

Nilsson N.J. (1965) *Learning Machines: Foundations of Trainable Pattern Classifying Systems*. New York: McGraw-Hill.

Oja E. (1983) *Subspace Methods of Pattern Recognition*. Letchworth: Research Studies Press.

Oliver J.J. (1993) *Decision graphs—an extension of decision trees*. Technical Report 173. Department of Computer Science, Monash University.

Oliver J.J. and Hand D.J. (1994) *Introduction to minimum encoding inference*. Technical report 94/205. Department of Computer Science, Monash University.

Oliver J.J. and Hand D.J. (1996) Averaging over decision trees. *Journal of Classification*, in press.

Olson H.F. and Belar H. (1956) Phonetic typewriter. *Journal of the Acoustical Society of America*, **28**, 1072–1081.

Overstreet G.A. and Bradley E.L. (1995) Applicability of generic linear scoring models in the USA Credit Union environment. Unpublished manuscript.

Palmieri G. and Sanna R. (1960) *Methodos*. **12**, No.48.

Pankhurst R.J. (1991) *Practical Taxonomic Computing*. Cambridge: Cambridge University Press.

Parzen E. (1962) On estimation of a probability density function and mode. *Annals of Mathematical Statistics*, **33**, 1065–1076.

Payne R.W. and Preece D.A. (1980) Identification keys and diagnostic tables (with discussion). *Journal of the Royal Statistical Society*, **143**, 253–292.

Peck R. and van Ness J. (1982) The use of shrinkage estimators in linear discriminant analysis. *IEEE Transactions on Pattern Analysis and Machine Intelligence*, **4**, 530–537.

Peck R., Jennings L.W., and Young D.M. (1988) A comparison of several biased estimators for improving the expected error rate of the sample quadratic discriminant function. *Journal of Statistical Computing and Simulation*, **29**, 143–156.

Perrone M.P.(1994) General averaging results for convex optimization. In *Proceedings of the 1993 Connectionist Models Summer School*, ed. M.C.Mozer, P.Smolensky, D.S.Tourtzby, J.Elman and A.Weigard Hillsdale, NJ: Lawrence Erlbaum, pp. 364–371.

Perrone M.P. and Cooper L.N. (1993) When networks disagree: ensemble methods for hybrid neural networks. In *Artificial Neural Networks for Speech and Vision*, ed. R.J.Mammone. London: Chapman and Hall, pp. 126–142.

Pickles A., Dunn G., and Vázquez-Barquero J.L. (1995) Reply to Hand (1995b). *Statistical Methods in Medical Research*, **4**, 263.

Piper J. (1992) Variability and bias in experimentally measured classifier error rates. *Pattern Recognition Letters*, **13**, 685–692.

Plaut D., Nowlan S., and Hinton G. (1986) Experiments on learning by back propagation. Technical report CMU-CS-86-26. Department of Computer Science, Carnegie Mellon University, Pittsburgh, PA.

Pruzek R.M. and Frederick B.C. (1978) Weighting predictors in linear models: alternatives to least squares and limitations of equal weights. *Psychological Bulletin*, **85**, 254–266.

Quenouille M.H. (1949) Approximate tests of correlation in time series. *Journal of the Royal Statistical Society, Series B*, **11**, 68–84.

Quinlan J.R. (1982) Semi-autonomous acquisition of pattern-based knowledge. In *Introductory Readings in Expert Systems*, ed. D.Michie. NewYork: Gordon and Breach, pp. 192–207.

Quinlan J.R. (1986) Induction of decision trees. *Machine Learning*, **1**, 81–106.

Quinlan J.R. (1992) Learning with continuous classes. In *Proceedings of the Fifth Australian Joint Conference on Artificial Intelligence*, ed A.Adams and L.Sterling. Singapore: World Scientific, pp. 343–348.

Quinlan J.R. (1993) *C4.5: Programs for Machine Learning*. San Mateo, CA: Morgan Kaufmann.

Quinlan J.R. and Rivest R.L. (1989) Inferring decision trees using the minimum description length principle. *Information and Computation*, **80**, 227–248.

Rabiner L.R. (1989) A tutorial on hidden Markov models and selected applications in speech recognition. *Proceedings of the IEEE*, **77**, 257–286.

Rabiner L. and Juang B.-H. (1993) *Fundamentals of Speech Recognition*. Englewood Cliffs, NJ: Prentice-Hall.

Rapoport A. (1975) Research paradigms for studying dynamic decision behaviour. In *Utility, Probability, and Human Decision Making*, ed. D.Wendt and C.A.J.Vlek. Dordrecht: D.Reidel.

Reichert A.K., Cho C.-C., and Wagner G.M. (1983) An examination of the conceptual issues involved in developing credit-scoring models. *Journal of Business and Economic Statistics*, **1**, 101–114.

Ridout M.S. (1988) An improved branch and bound algorithm for feature subset selection. *Applied Statistics*, **37**, 139–147.

Ringrose T.J. and Krzanowski W.J. (1991) Simulation study of confidence regions for canonical variate analysis. *Statistics and Computing*, **1**, 41–46.

Ripley B.D. (1996) *Pattern Recognition and Neural Networks*. Cambridge: Cambridge University Press.

Rissanen J. (1987) Stochastic complexity (with discussion). *Journal of the Royal Statistical Society, Series B*, **49**, 223–239.

Rissanen J. (1989) *Stochastic Complexity in Statistical Inquiry*. Singapore: World Scientific.

Rodriguez A.F. (1988) Admissibility and unbiasedness of the ridge classification rules for two normal populations with equal covariance matrices. *Statistics*, **19**, 383–388.

Rohwer R. (1990) The 'moving targets' training algorithm. In *Advances in Neural Information Processing Systems II*, ed. D.S.Touretzky. San Mateo, CA: Morgan Kaufmann, pp. 558–565.

Rosenberg E. and Gleit A. (1994) Quantitative methods in credit management: a survey. *Operations Research*, **42**, 589–613.

Rosenblatt F. (1962) *Principles of Neurodynamics: Perceptrons and the Theory of Brain Mechanisms*. Washington, D.C.: Spartan Books.

Rosenblatt M. (1956) Remarks on some nonparametric estimates of a density function. *Annals of Mathematical Statistics*, **27**, 832–837.

Rounds E. (1980) A combined non-parametric approach to feature selection and binary decision tree design. *Pattern Recognition*, **12**, 313–317.

Rumelhart D.E., Hinton G.E., and Williams R.J.(1986) Learning internal representations by error propagation. In *Parallel Distributed Processing: Explorations in the Microstructures of Cognition, Vol. 1: Foundations*, ed. D.E.Rumelhart and J.L.McClelland, Cambridge, MA: MIT Press, pp. 318–362.

Safavian S.R. and Landgrebe D. (1991) A survey of decision tree classifier methodology. *IEEE Transactions on Systems, Man, and Cybernetics*, **21**, 660–674.

Sanders F. (1963) On subjective probability forecasting. *Journal of Applied Meteorology*, **2**, 191–201.

Sanders F. (1973) Skill in forecasting daily temperature and precipitation: some experimental results. *Bulletin of the American Meteorological Society*, **54**, 1171–1179.

Schaffer C. (1994) Cross-validation, stacking and bi-level stacking: meta-methods for classification and learning. In *Selecting Models from Data: AI and Statistics IV*, ed. P.Cheeseman and R.W.Oldford. New York: Springer-Verlag, pp. 51–59.

Schuerman J. and Doster D. (1984) A decision-theoretic approach in hierarchical classifier design. *Pattern Recognition*, **17**, 359–369.

Schwarz G. (1978) Estimating the dimension of a model. *Annals of Statistics*, **6**, 461–464.

Scott D.W. (1992) *Multivariate Density Estimation*. New York: John Wiley.

Seung H.S., Sompolinsky H., and Tishby N. (1992) Statistical mechanics of learning from examples. *Physical Review A*, **45**, 6056–6091.

Shepherd B., Piper J., and Rutovitz D. (1987) Comparison of ACLS and classical linear methods in a biological application. In *Machine Intelligence*, Vol. 11, ed. J.E.Hayes, D.Michie, and J.Richards. Oxford: Oxford University Press, pp. 423–434.

Short R.D. and Fukunaga K. (1981) Optimal distance measure for nearest neighbour classification. *IEEE Transactions on Information Theory*, **27**, 622–627.

Silverman B.W. (1986) *Density Estimation for Statistics and Data Analysis*. London: Chapman and Hall.

Smith C.A.B. (1947) Some examples of discrimination. *Annals of Eugenics*, **13**, 272–282.

Snapinn S.M. and Knoke J.D. (1985) An evaluation of smoothed classification error rate estimators. *Technometrics*, **27**, 199–206.

Snapinn S.M. and Knoke J.D. (1988) Bootstrapped and smoothed classification error rate estimators. *Communications in Statistics, Simulation*, **17**, 1135–1153.

Snapinn S.M. and Knoke J.D. (1989) Estimation of error rates in discriminant analysis with selection of variables. *Biometrics*, **45**, 289–299.

Solla S.A., Levin E., and Fleisher M. (1988) Accelerated learning in layered neural networks. *Complex Systems*, **2**, 625–639.

Sonquist J.N. (1970) *Multivariate model building*. Institute for Social Research, University of Michigan.

Späth H. (1985) *Cluster Dissection and Analysis*. Chichester: Ellis Horwood.

Speed T.P. and Kiiveri H.T. (1986) Gaussian Markov distributions over finite graphs. *Annals of Statistics*, **14**, 138–150.

Stanfill C and Waltz D. (1986) Toward memory-based reasoning. *Communications of the ACM*, **29**, 1213–1228.

Stone M. and Brooks R.J. (1990) Continuum regression: cross-validated sequentially constructed prediction embracing ordinary least squares, partial least squares and principal components regression (with discussion). *Journal of the Royal Statistical Society, Series B*, **52**, 237–269.

Stone M. and Jonathan P. (1993) Statistical thinking and technique for QSAR and related studies. Part I: general theory. *Journal of Chemometrics*, **7**, 455–475.

Stone M. and Jonathan P. (1994) Statistical thinking and technique for QSAR and related studies. Part II: Specific methods. *Journal of Chemometrics*, **8**, 1–20.

Sturt E. (1981) Computerized construction in Fortran of a discriminant function for categorical data. *Applied Statistics*, **30**, 213–222.

Suen C.Y., Nadal C., Legault R., Mai T.A., and Lam L. (1992) Computer recognition of unconstrained handwritten numerals. *Proceedings of the IEEE*, **80**, 1162–1180.

Taylor P.C. and Silverman B.W. (1993) Block diagrams and splitting criteria for classification trees. *Statistics and Computing*, **3**, 147–161.

Timmers T. (1987) *Pattern Recognition of Cytological Specimens*. PhD thesis. University of Amsterdam.

Titterington D.M., Murray G.D., Murray L.S., Spiegelhalter D.J., Skene A.M., Habbema J.D.F., and Gelpke G.J. (1981) Comparison of discrimination techniques applied to a complex data set of head injured patients. *Journal of the Royal Statistical Society, Series A*, **144**, 145–175.

Todeschini R. (1989) k-nearest neighbour method: the influence of data transformations and metrics. *Chemometrics Intell. Laboratory Systems*, **6**, 213–220.

Toussaint G.T. (1974) Bibliography on estimation of misclassification. *IEEE Transactions on Information Theory*, **20**, 472–479.

Toussaint G.T. (1975) An efficient method for estimating the probability of misclassification applied to a problem in medical diagnosis. *Computers in Biology and Medicine*, **4**, 269–278.

Valiant L.G. (1984) A theory of the learnable. *Communications of the ACM*, **27**, 1134–1142.

Velilla S. and Barrio J.A. (1994) A discriminant rule under transformation. *Technometrics*, **36**, 348–353.

Venables W.N. and Ripley B.D. (1994) *Modern Applied Statistics with S-Plus*. New York: Springer-Verlag.

von Winterfeldt D. and Edwards W. (1982) Costs and payoffs in perceptual research. *Psychological Bulletin*, **91**, 609–622.

Waibel A. and Lee K.-F. (1990) *Readings in Speech Recognition*. San Mateo, CA: Morgan Kaufmann.

Wainer H. (1976) Estimating coefficients in linear models: it don't make no nevermind. *Psychological Bulletin*, **83**, 213–217.

Wainer H. (1978) On the sensitivity of regression and regressors. *Psychological Bulletin*, **85**, 267–273.

Wallace C.S. and Freeman P.R. (1987) Estimation and inference by compact encoding (with discussion). *Journal of the Royal Statistical Society, Series B*, **49**, 240–265.

Wallace C.S. and Patrick J.D. (1993) Coding decision trees. *Machine Learning*, **11**, 7–22.

Wand M.P. and Jones M.J. (1995) *Kernel Smoothing*. London: Chapman and Hall.

Watanabe S. (1965) Karhunen-Loeve expansion and factor analysis. *Transactions of the 4th Prague Conference on Information Theory, Statistical Decision Functions, and Random Processes*. Prague, 1965, pp. 635–660. Reprinted in *Pattern Recognition—An Introduction and Foundation*, ed. J.Sklansky (1973). Dowden: Hutchinson and Ross.

Watanabe S. (1970) Feature compression. In *Advances in Information Systems Science*, Vol.3, ed. J.T.Tou. New York: Plenum Press.

Watkin T.L.H., Rau A., and Biehl M. (1993) The statistical mechanics of learning a rule. *Reviews of Modern Physics*, **65**, 499–556.

Wermuth N. (1980) Linear recursive systems, covariance selection and path analysis. *Journal of the American Statistical Association*, **75**, 963–972.

Wermuth N. (1991) *On block-recursive linear regression equations*. Research report. University of Mainz Psychological Institute.

Whittaker J. (1990) *Graphical Models in Applied Multivariate Statistics*. Chichester: John Wiley.

Widrow B. (1962) Generalization and information storage in networks of adaline 'neurons'. In *Self-Organizing Systems 1962, Chicago*, ed. M.C.Yovits, G.T.Jacobi, and G.D.Goldstein. Washington: Spartan Books, pp. 435–461.

Widrow B. and Hoff M.E. (1960) Adaptive switching circuits. In *1960 IRE WSECON Convention Record*. Part 4, 96–104. New York: IRE. Reprinted in *Neurocomputing: Foundations of Research*, ed. J.A. Anderson and E.Rosenfeld. Cambridge, MA: MIT Press.

Wieand S., Gail M.H., James B.R., and James K.L. (1989) A family of nonparamteric statistics for comparing diagnostic markers with paired or unpaired data. *Biometrika*, **76**, 585–592.

Williams P., Hand D.J., and Tarnopolsky A. (1982) The problem of screening for uncommon disorders—a comment on the eating attitudes test. *Psychological Medicine*, **12**, 431–434.

Wold H. (1966) Estimation of principal components and related models by iterative least squares. In *Multivariate Analysis*, ed. P.R.Krishnaiah. New York: Academic Press.

Wold H. (1985) Partial least squares. In *Encyclopedia of Statistical Sciences*, ed. S.Kotz and N.Johnson. New York: John Wiley, pp. 581–591.

Wold S. (1976) Pattern recognition by means of disjoint principal components models. *Pattern Recognition*, **8**, 127–139.

Wolpert D.H. (1992) Stacked generalization. *Neural Networks*, **5**, 241–259.

Xu L., Krzyzak A., and Suen C.Y. (1992) Methods of combining multiple classifiers and their applications to handwriting recognition. *IEEE Transactions on Pattern Analysis and Machine Intelligence*, **22**, 418–435.

Yates F. (1982) External correspondence: decompositions of the mean probability score. *Organizational Behaviour and Human Performance*, **30**, 132–156.

Zweig M.H. and Campbell G. (1993) Receiver-operating characteristic (ROC) plots. *Clinical Chemistry*, **29**, 561–577.

Index

Index compiled by Geoffrey C. Jones

WILEY SERIES IN PROBABILITY AND STATISTICS

ESTABLISHED BY WALTER A. SHEWHART AND SAMUEL S. WILKS
Editors
*Vic Barnett, Ralph A. Bradley, Nicholas I. Fisher, J.B. Kadane,
David G. Kendall, David W. Scott, Adrian F. M. Smith, Jozef L. Teugels,
Geoffrey S. Watson*

Probability and Statistics

*Now available in a lower priced paperback edition in the Wiley Classics Library

LAMPERTI • Probability: A Survey of the Mathematical Theory, *Second Edition*
LARSON • Introduction to Probability Theory and Statistical Inference, *Third Edition*
LESSLER and KALSBEEK • Nonsampling Error in Surveys
LINDVALL • Lectures on the Coupling Method
McLACHLAN • Discriminant Analysis and Statistical Pattern Recognition
McLACHLAN and KRISHNAN • The EM Algorithm
McNEIL • Epidemiological Research Methods
MANTON, WOODBURY and TOLLEY • Statistical Applications Using Fuzzy Sets
MARDIA • The Art of Statistical Science: A Tribute to G.S. Watson
MARDIA and DRYDEN • Statistical Analysis of Shape
MOLCHANOV • Statistics of the Boolean Model for Practitioners and Mathematicians
MORGENTHALER and TUKEY • Configural Polysampling: A Route to Practical Robustness
MUIRHEAD • Aspects of Multivariate Statistical Theory
OLIVER and SMITH • Inference Diagrams, Belief Nets and Decision Analysis
*PARZEN • Modern Probability Theory and Its Applications
PRESS • Bayesian Statistics: Principles, Models, and Applications
PUKELSHEIM • Optimal Experimental Design
PURI and SEN • Nonparametric Methods in General Linear Models
PURI, VILAPLANA and WERTZ • New Perspectives in Theoretical and Applied Statistics
RAO • Asymptotic Theory of Statistical Inference
RAO • Linear Statistical Inference and Its Applications, *Second Edition*
RAO and SHANBHAG • Choquet-Deny Type Functional Equations and Applications to Stochastic Models
RENCHER • Methods of Multivariate Analysis
ROBERTSON, WRIGHT and DYKSTRA • Order Restricted Statistical Inference
ROGERS and WILLIAMS • Diffusions, Markov Processes, and Martingales, Volume I: Foundations,
        Second Edition, Volume II: Itô Calculus
ROHATGI • An Introduction to Probability Theory and Mathematical Statistics
ROSS • Stochastic Processes
RUBINSTEIN • Simulation and the Monte Carlo Method
RUBINSTEIN and SHAPIRO • Discrete Event Systems: Sensitivity Analysis and Stochastic Optimization
        by the Score Function Method
RUZSA and SZEKELY • Algebraic Probability Theory
SCHEFFE • The Analysis of Variance
SEBER • Linear Regression Analysis
SEBER • Multivariate Observations
SEBER and WILD • Nonlinear Regression
SERFLING • Approximation Theorems of Mathematical Statistics
SHORACK and WELLNER • Empirical Processes with Applications to Statistics
SMALL and McLEISH • Hilbert Space Methods in Probability and Statistical Inference
STAPLETON • Linear Statistical Models
STAUDTE and SHEATHER • Robust Estimation and Testing
STOYANOV • Counterexamples in Probability, *Second Edition*
STYAN • The Collected Papers of T.W. Anderson 1943–1985
TANAKA • Time Series Analysis. Nonstationary and Noninvertible Distribution Theory
THOMPSON and SEBER • Adaptive Sampling
WELSH • Aspects of Statistical Inference
WHITTAKER • Graphic Models in Applied Multivariate Statistics
WILLIAMS • Diffusions, Markov Processes, and Martingales, Volume 1. *Second Edition*
YANG • The Construction Theory of Denumerable Markov Processes

Applied Probability and Statistics
ABRAHAM and LEDOLTER • Statistical Methods for Forecasting
AGRESTI • Analysis of Ordinal Categorical Data
AGRESTI • Categorical Data Analysis
AGRESTI • An Introduction to Categorical Data Analysis

*Now available in a lower priced paperback edition in the Wiley Classics Library

ANDERSON and LOYNES • The Teaching of Practical Statistics
ANDERSON, AUQUIER, HAUCH, OAKES, VANDAELE and WEISBERG • Statistical Methods for Comparative Studies
ARMITAGE and DAVID (editors) • Advances in Biometry
*ARTHANARI and DODGE • Mathematical Programming in Statistics
ASMUSSEN • Applied Probability and Queues
*BAILEY • The Elements of Stochastic Processes with Applications to the Natural Sciences
BARNETT and LEWIS • Outliers in Statistical Data, *Third Edition*
BARTHOLOMEW, FORBES, and McLEAN • Statistical Techniques for Manpower Planning, *Second Edition*
BATES and WATTS • Nonlinear Regression Analysis and Its Applications
BECHOFER, SANTNER and GOLDSMAN • Design and Analysis of Experiments for Statistical Selection, Screening and Multiple Comparisons
BELSLEY • Conditioning Diagnostics: Collinearity and Weak Data in Regression
BELSLEY, KUH and WELSCH • Regression Diagnostics: Identifying Influential Data and Sources of Collinearity •
BERNARDO and SMITH • Bayesian Theory
BERRY, CHALONER and GEWEKE • Bayesian Analysis in Statistics and Econometrics Essays in Honor of Arnold Zellner
BHAT • Elements of Applied Stochastic Processes, *Second Edition*
BHATTACHARYA and WAYMIRE • Stochastic Processes with Applications
BIEMER, GROVES, LYBERG, MATHIOWETZ and SUDMAN • Measurement Errors in Surveys
BIRKES and DODGE • Alternative Methods of Regression
BLOOMFIELD • Fourier Analysis of Time Series: An Introduction
BOLLEN • Structural Equations with Latent Variables
BOULEAU • Numerical Methods for Stochastic Processes
BOX • R.A.Fisher, the Life of a Scientist
BOX and DRAPER • Empirical Model-Building and Response Surfaces
BOX and DRAPER • Evolutionary Operation: A Statistical Method for Process Improvement
BOX, HUNTER and HUNTER • Statistics for Experimenters: An Introduction to Design, Data Analysis, and Model Building
BROWN and HOLLANDER • Statistics: A Biomedical Introduction
BUCKLEW • Large Deviation Techniques in Decision, Simulation, and Estimation
BUNKE and BUNKE • Non-linear Regression, Functional Relations and Robust Methods: Statistical Methods of Model Building
CHATTERJEE and HADI • Sensitivity Analysis in Linear Regression
CHATTERJEE and PRICE • Regression Analysis by Example, *Second Edition*
CLARKE and DISNEY • Probability and Random Processes: A First Course with Applications, *Second Edition*
COCHRAN • Sampling Techniques, *Third Edition*
*COCHRAN and COX • Experimental Designs, *Second Edition*
CONOVER • Practical Nonparametric Statistics, *Second Edition*
CORNELL • Experiments with Mixtures, Designs, Models, and the Analysis of Mixture Data, *Second Edition*
COX • A Handbook of Introductory Statistical Methods
*COX • Planning of Experiments
COX, BINDER, CHINNAPPA, CHRISTIANSON, COLLEDGE, and KOTT • Business Survey Methods
CRESSIE • Statistics for Spatial Data, *Revised Edition*
DANIEL • Applications of Statistics to Industrial Experimentation
DANIEL • Biostatistics: A Foundation for Analysis in the Health Sciences, *Sixth Edition*
DANIEL Fitting Equations into Data: Computer Analysis of Multifactor Data, *Second Edition*
DAVID • Order Statistics, *Second Edition*
*DEGROOT, FIENBERG and KADANE • Statistics and the Law
*DEMING • Sample Design in Business Research
DILLON and GOLDSTEIN • Multivariate Analysis: Methods and Applications
DOWDY and WEARDEN • Statistics for Research, *Second Edition*
DRAPER and SMITH • Applied Regression Analysis, *Second Edition*
DUNN • Basic Statistics: A Primer for the Biomedical Sciences, *Second Edition*

*Now available in a lower priced paperback edition in the Wiley Classics Library

DUNN and CLARK • Applied Statistics: Analysis of Variance and Regression, *Second Edition*
DUPUIS • Large Deviations
ELANDT-JOHNSON and JOHNSON • Survival Models and Data Analysis
EVANS, PEACOCK and HASTINGS • Statistical Distributions, *Second Edition*
FISHER and VAN BELLE • Biostatistics: A Methodology for the Health Sciences
FLEISS • The Design and Analysis of Clinical Experiments
FLEISS • Statistical Methods for Rates and Proportion, *Second Edition*
FLEMING and HARRINGTON • Counting Processes and Survival Analysis
FLURY • Common Principal Components and Related Multivariate Models
GALLANT • Nonlinear Statistical Models
GHOSH • Estimation
GLASSERMAN and YAO • Monotone Structure in Discrete-Event Systems
GNANADESIKAN • Analysis, *Second Edition*
GOLDSTEIN and LEWIS • Assessment: Problems, Developments and Statistical Issues
GOLDSTEIN and WOOFF • Bayes Linear Statistics
GREENWOOD and NIKULIN • A Guide to Chi-squared Testing
GROSS and HARRIS • Fundamentals of Queuing Theory, *Second Edition*
GROVES • Survey Errors and Survey Costs
GROVES, BIEMER, LYBERG, MASSEY, NICHOLLS and WAKSBERG • Telephone Survey
      Methodology
HAHN • and MEEKER • Statistical Intervals: A Guide for Practitioners
HAND • Construction and Assessment of Classification Rules
HAND • Discrimination and Classification
*HANSEN, HURWITZ and MADOW • Sample Survey Methods and Theory, Volume 1: Methods and
      Applications
*HANSEN, HURWITZ and MADOW • Sample Survey Methods and Theory, Volume II: Theory
HEIBERGER • Computation for the Analysis of Designed Experiments
HELLER • MACSYMA for Statisticians
HINKELMAN and KEMPTHORNE • Design and Analysis of Experiments, Volume 1: Introduction to
      Experimental Design
HOAGLIN, MOSTELLER and TUKEY • Exploratory Approach to Analysis of Variance
HOAGLIN, MOSTELLER and TUKEY • Exploring Data Tables, Trends and Shapes
HOAGLIN, MOSTELLER and TUKEY • Understanding Robust and Exploratory Data Analysis
HOCHBERG and TAMHANE • Multiple Comparison Procedures
HOCKING • Methods and Applications of Linear Models: Regression and the Analysis of Variance
HOEL • Elementary Statistics, *Fifth Edition*
HOGG and KLUGMAN • Loss Distributions
HOLLANDER and WOLFE • Nonparametric Statistical Methods
HOSMER and LEMESHOW • Applied Logistic Regression
HØYLAND and RAUSAND • System Reliability Theory: Models and Statistical Methods
HUBERTY • Applied Discriminant Analysis
IMAN and CONOVER • Modern Business Statistics
JACKSON • A User's Guide to Principle Components
JOHN • Statistical Methods in Engineering and Quality Assurance
JOHNSON • Multivariate Statistical Simulation
JOHNSON & KOTZ • Distributions in Statistics
                     • Continuous Multivariate Distributions
JOHNSON, KOTZ and BALAKRISHNAN • Continuous Univariate Distributions, Volume 1, *Second
      Edition; Volume 2, Second Edition*
JOHNSON, KOTZ, and BALAKRISHNAN • Discrete Multivariate Distributions
JOHNSON, KOTZ and KEMP • Univariate Discrete Distribution, *Second Edition*
JUDGE, GRIFFITHS, HILL, LÜTKEPOHL, and LEE • The Theory and Practice of Econometrics, *Second
      Edition*
JUDGE, HILL, GRIFFITHS, LÜTKEPOHL, and LEE • Introduction to the Theory and Practice of
      Econometrics, *Second Edition*
JURECKOVÁ and SEN • Robust Statistical Procedures: Asymptotics and Interrelations
KADANE • Bayesian Methods and Ethics in a Clinical Trial Design
KADANE and SCHUM • A Probabilistic Analysis of the Sacco and Vanzetti Evidence

*Now available in a lower priced paperback edition in the Wiley Classics Library

KALBFLEISCH and PRENTICE • The Statistical Analysis of Failure Time Data
KASPRZYK, DUNCAN, KALTON and SINGH • Panel Surveys
KHURI • Advanced Calculus with Applications in Statistics
KISH • Statistical Design for Research
*KISH • Survey Sampling
KOTZ • Personalities
KOVALENKO, KUZNETZOV and PEGG • Mathematical Theory of Reliability of Time-dependent
    Systems with Practical Applications
LAD • Operational Subjective Statistical Methods: A Mathematical, Philosophical and Historical
    Introduction
LANGE, RYAN, BILLARD, BRILLINGER, CONQUEST, and GREENHOUSE • Case Studies in
    Biometry
LAWLESS • Statistical Models and Methods for Lifetime Data
LEE • Statistical Methods for Survival Data Analysis, *Second Edition*
LePAGE and BILLARD • Exploring the Limits of Bootstrap
LESSLER and KALSBEEK • Nonsampling Error in Surveys
LEVY and • LEMESHOW • Sampling of Populations: Methods and Applications
LINHART and ZUCCHINI • Model Selection
LITTLE and RUBIN Statistical Analysis with Missing Data
LYBERG • Survey Measurement
McLACHLAN • Discriminant Analysis and Statistical Pattern Recognition
McLACHLAN and KRISHNAN • The EM Algorithm and Extensions
McNEIL • Epidemiological Research Methods
MAGNUS and NEUDECKER • Matrix Differential Calculus with Applications in Statistics and
    Econometrics
MALLER and ZHOU • Survival Analysis with Long Term Survivors
MALLOWS • Design, Data, and Analysis by Some Friends of Cuthbert Daniel
MANN, SCHAFER, and SINPURWALLA • Methods for Statistical Analysis of Reliability and Life Data
MASON, GUNST, and HESS • Statistical Design and Analysis of Experiments with Applications to
    Engineering and Science
MILLER • Survival Analysis
MONTGOMERY and MYERS • Response Surface Methodology: Process and Product in Optimization
    Using Designed Experiments
MONTGOMERY and PECK • Introduction to Linear Regression Analysis, *Second Edition*
MORGENTHALER and TUKEY • Configural Polysampling
MYERS and MONTGOMERY • Response Surface Methodology
NELSON • Accelerated Testing, Statistical Models, Test Plans, and Data Analyses
NELSON • Applied Life Data Analysis
OCHI • Applied Probability and Stochastic Processed in Engineering and Physical Sciences
OKABE, BOOTS, and SUGIHARA • Spatial Tesselations: Concepts and Applications of Voronoi
    Diagrams
PANKRATZ • Forecasting with Dynamic Regression Models
PANKRATZ • Forecasting with Univariate Box-Jenkins Models: Concepts and Cases
PORT • Theoretical Probability for Applications
PUKELSHEIM • Optimal Design of Experiments
PUTERMAN • Markov Decision Processes: Discrete Stochastic Dynamic Programming
RACHEV • Probability Metrics and the Stability of Stochastic Models
RADHAKRISHNA RAO and SHANBHAG • Choquet-Deny Type Functional Equations with Applications
    to Stochastic Models
RÉNYI • A Diary on Information Theory
RIPLEY • Spatial Statistics
RIPLEY • Stochastic Simulation
ROSS • Introduction to Probability and Statistics for Engineers and Scientists
ROUSSEEUW and LEROY • Robust Regression and Outlier Detection
RUBIN • Multiple Imputation for Nonresponse in Surveys
RUBINSTEIN and SHAPIRO Discrete Event Systems: Sensitivity Analysis and Stochastic Optimization by
    the Score
RYAN • Modern Regression Methods

RYAN • Statistical Methods for Quality Improvement
SCHOTT • Matrix
SCOTT • Multivariate Density Estimation: Theory, Practice, and Visualization
SEARLE • Linear Models
SEARLE • Linear Models for Unbalanced Data
SEARLE • Matrix Algebra Useful for Statistics
SEARLE, CASELLA and McCULLOCH • Variance Components
SKINNER, HOLT, and SMITH • Analysis of Complex Surveys
STOYAN, KENDALL, and MECKE • Stochastic Geometry and Its Applications, *Second Edition*
STOYAN and STOYAN • Fractals, Random Shapes and Point Fields: Methods of Geometrical Statistics
THOMPSON • Empirical Model Building
THOMPSON • Sampling
TIERNEY • LISP-STAT: An Object-Oriented Environment for Statistical Computing and Dynamic
    Graphics
TIJMS • Stochastic Models: An Algorithmic Approach
TITTERINGTON, SMITH and MARKOV • Statistical Analysis of Finite Mixture Distributions
UPTON and FINGLETON • Spatial Data Analysis by Example, Volume 1: Point Pattern and Quantitative
    Data
UPTON and FINGLETON • Spatial Data Analysis by Example, Volume II: Categorical and
    Directional Data
VAN RIJKEVORSEL and DE LEEUW • Component and Correspondence Analysis
WEISBERG • Applied Linear Regression, *Second Edition*
WESTFALL and YOUNG • Resampling-Based Multiple Testing: Examples and Methods for *p*-Value
    Adjustment
WHITTLE • Optimization Over Time: Dynamic Programming and Stochastic Control, Volume 1 and
    Volume II
WHITTLE • Systems in Stochastic Equilibrium
WONNACOTT and WONNACOTT • Econometrics, *Second Edition*
WONNACOTT and WONNACOTT • Introductory Statistics, *Fifth Edition*
WONNACOTT and WONNACOTT • Introductory Statistics for Business and Economics, *Fourth Edition*
WOODING • Planning Pharmaceutical Clinical Trials: Basic Statistical Principles
WOOLSON • Statistical Methods for the Analysis of Biomedical Data
*ZELLNER • An Introduction to Bayesian Inference in Econometrics

Tracts on Probability and Statistics
BILLINGSLEY • Convergence of Probability Measures
KELLY • Reversibility and Stochastic Networks

*Now available in a lower priced paperback edition in the Wiley Classics Library